Neural Search – From Prototype to Production with Jina

Build deep learning–powered search systems that you can deploy and manage with ease

Jina AI

Bo Wang

Cristian Mitroi

Feng Wang

Shubham Saboo

Susana Guzmán

BIRMINGHAM—MUMBAI

Neural Search – From Prototype to Production with Jina

Group Product Manager: Kunal Chaudhari

Publishing Product Manager: Dhruv Jagdish Kataria

Senior Editor: Nisha Cleetus

Content Development Editor: Nithya Sadanandan

Technical Editor: Pradeep Sahu

Copy Editor: Safis Editing

Project Coordinator: Ajesh Devavaram

Proofreader: Safis Editing

Indexer: Tejal Daruwale Soni

Production Designer: Shankar Kalbhor

Marketing Coordinator: Abeer Riyaz Dawe

Business Development Executive: Surya Srivastav

First published: September 2022

Production reference: 1160922

Published by Packt Publishing Ltd.
Livery Place
35 Livery Street
Birmingham
B3 2PB, UK.

ISBN 978-1-80181-682-3

www.packt.com

Contributors

About the authors

Bo Wang is a machine learning engineer at Jina AI. He has a background in computer science, especially interested in the field of information retrieval. In the past years, he has been conducting research and engineering work on search intent classification, search result diversification, content-based image retrieval, and neural information retrieval. At Jina AI, Bo is working on developing a platform for automatically improving search quality with deep learning. In his spare time, he likes to play with his cats, watch anime, and play mobile games.

This book is a joint effort between Jina AI and Packt Publishing. I appreciate everyone who organized and participated in writing this unique book. Thank you, Bing, for giving us such a great opportunity to share our learnings with others, and thanks to all the co-authors, Shubham, Susana, Feng, Alex Cureton-Griffiths, and Cristian, for their hard work. I would also like to share my gratitude to Zizhen Wang for helping me with Chapter 1 and Chapter 2 of the book.

I feel honored to have worked with a professional editing team, including Nithya Sadanandan and Prajakta Naik. Your feedback always helped me better convey my thoughts with great clarity.

Cristian Mitroi is a machine learning engineer with a wide breadth of experience in full stack, from infrastructure to model iteration and deployment. His background is based in linguistics, which led to him focusing on NLP. He also enjoys, and has experience in, teaching and interacting with the community, and has given workshops at various events. In his spare time, he performs improv comedy and organizes too many pen-and-paper role-playing games.

I would like to thank the entire Jina AI team for providing a great working environment and being massively supportive and helpful. A special thank you goes to Han, for laying the foundation. I would also like to thank my mentor, Max, for all the memorable insights. Last but not least, I want to thank my supportive girlfriend, for bearing with my nerdy monologues.

Feng Wang is a machine learning engineer at Jina AI. He received his Ph.D. from the department of computer science at the Hong Kong Baptist University in 2018. He has been a full-time R&D engineer for the past few years, and his interests include data mining and artificial intelligence, with a particular focus on natural language processing, multi-modal representation learning, and recommender systems. In his spare time, he likes climbing, hiking, and playing mobile games.

I would like to thank everyone who helped me with the writing process. It's been a great honor to have this chance to share my learnings of neural search with the world. I am grateful to have worked with my brilliant co-authors, Bo, Susana, Alex Cureton-Griffiths, Shubham, and Christian. Thanks to our professional editors, Nithya Sadanandan and Prajakta Naik. Their work has been a great help in making this book available to the world. I would also like to thank my wife, Ting Wang, for being there for me. She acted as a catalyst for me to write this book.

Shubham Saboo has taken on multiple roles, from a data scientist to an AI evangelist, at renowned firms across the globe, where he was involved in building organization-wide data strategies and technology infrastructure to create and scale data teams from scratch. His work as an AI evangelist has led him to build communities and reach out to a broader audience to foster the exchange of ideas and thoughts in the burgeoning field of AI. As part of his passion for learning new things and sharing knowledge with the community, he writes technical blogs on the advancements in AI and its economic implications. In his spare time, you can find him traveling the world, which enables him to immerse himself in different cultures and refine his worldview.

I would like to acknowledge Han Xiao, for coming up with the Jina framework to make neural search accessible to all, and Bing He, for giving me the opportunity to write this book. I'm grateful to have worked with my brilliant co-authors, Bo, Susana, Alex Cureton-Griffiths, Feng, and Christian, who helped me throughout the writing process and deepened my understanding of neural search. Huge thanks to the editors, Nithya Sadanandan and Prajakta Naik, who did a great job at shaping the book into its final form.

I'd also like to thank my mom, Gayatri, who always believed in me irrespective of the odds. My love, Gargi, for being there for me at every step and making this journey of writing the book a blissful experience.

Susana Guzmán is the product manager at Jina AI. She has a background in computer science and for several years was working at different firms as a software developer with a focus on computer vision, working with both C++ and Python. She has a big interest in open source, which was what led her to Jina, where she started as a software engineer for 1 year until she got a clear overview of the product, which made her make the switch from engineering to PM. In her spare time, she likes to cook food from different cuisines around the world, looking for her new favorite dish.

I want to take a moment to thank all the people who helped make this book possible. To Han, Nan, and Bing, for starting this amazing journey of neural search and giving me the opportunity to grow with them. My team helped me every time I had questions, ranging from technical to writing, especially Shubham, Bo, Alex Cureton-Griffiths, Feng, and Cristian. You are amazing colleagues and I'm grateful to work with you. To all the Packt team, who were always available and gave so many valuable inputs. Thanks a lot. And lastly, though not least important, thanks to coffee. This wouldn't have been possible without you either.

About the reviewer

Dr. Ibrahim Haddad is vice president of strategic programs at the Linux Foundation and the general manager of LF AI & Data. He is focused on collaborating with the largest technology companies and open source projects, facilitating a vendor-neutral environment for advancing the open source platform, and empowering generations of innovators by providing a neutral, trusted hub for developers to code, manage, and scale technology projects. Throughout his career, Haddad has held technology and portfolio management roles at Ericsson Research, Open Source Development Labs, Motorola, Palm, Hewlett-Packard, and Samsung Research. He graduated with Honors from Concordia University (Montréal, Canada) with a Ph.D. in computer science.

Table of Contents

3

Part 2: Introduction to Jina Fundamentals

4

5

Part 3: How to Use Jina for Neural Search

6

7

Preface

We live in the digital era, and creating data becomes easier every day. According to Forbes, we generate 2.5 quintillion bytes of data each day, and this data comes from all types of sources. It can be pictures on Instagram, voice messages on Telegram, videos, text, or even a combination of all of them. This was definitely not the case when the internet was just starting. And yet, despite the obvious difference in the easiness of data creation before and now, we keep using the same search techniques. Despite having an incredible boom in data generation, we haven't updated our search techniques that much.

Neural search is the approach to changing that. It takes advantage of the **machine learning (ML)** era that we live in right now and uses the latest AI research to deliver novel search techniques. However, despite this being a good technique, it presents a lot of challenges. Neural search is a whole new concept, which means the knowledge and techniques needed are also new. To effectively deploy a neural search application, the engineers in charge of it need to have a plethora of skills, from engineering to Dev-Ops, and an ML background.

This is the problem Jina AI wants to solve. Jina is an open source solution that is designed to democratize AI and neural search, making it easier for many developers to have a full end-to-end neural search application without having to have a background in ML, the cloud, and backend engineering. It's designed with Python developers in mind to help them unlock the full potential of the latest neural search techniques. In this book, we will explore the basics of search, from traditional to neural search. With this knowledge, we will work through hands-on examples of creating a fully fledged neural application.

Who this book is for

If you are an ML, **deep learning (DL)**, or AI engineer interested in building a search system of any kind (text, QA, image, audio, PDF, 3D models, and so on) using modern software architecture, this book is for you. This book is also excellent for Python engineers who are interested in building a search system (of any kind) using state-of-the-art DL techniques.

What this book covers

Chapter 1, Neural Networks for Neural Search, will cover what neural search is and what it is used for. You'll see an overview of its practical applications, the challenges it poses, and how to overcome them.

Chapter 2, Introducing Foundations of Vector Representation, will cover the concept of vectors in ML. It will also introduce you to common search algorithms built on top of vector representations and their strengths and weaknesses.

Chapter 3, *System Design and Engineering Challenges*, will cover the basics of designing a search system. In this chapter, you will learn the core concepts associated with search, such as indexing and querying, and how to use this to save and retrieve information.

Chapter 4, *Learning Jina's Basics*, will cover the steps required to implement your own search engine. In this chapter, you will learn the core concepts of Jina in detail, along with their overall design and how they connect with each other.

Chapter 5, *Multiple Search Modalities*, will introduce you to the concept of multi-modal and cross-modal search, where you can combine multiple modalities, such as text, image, audio, and video, to build state-of-the-art search systems.

Chapter 6, *Basic Practical Examples with Jina*, will cover beginner-friendly real-world applications of Jina's neural search framework by building on the concepts learned in the previous chapters.

Chapter 7, *Exploring Advanced Use Cases of Jina*, will cover the advanced applications of Jina's neural search framework by using the concepts learned in the previous chapters. It will focus on explaining challenging concepts in neural search using real-world examples.

To get the most out of this book

To be able to make the most of this book, you are expected to write programs with Python and have knowledge of ML and DL. It would be even better to have a basic understanding of information retrieval and the search problem.

Software/hardware covered in the book	Operating system requirements
Python 3.7	Windows with WSL, macOS, or Linux
Jina 3.7	
DocArray 0.13	

If you are using the digital version of this book, we advise you to type the code yourself or access the code from the book's GitHub repository (a link is available in the next section). Doing so will help you avoid any potential errors related to the copying and pasting of code.

Download the example code files

You can download the example code files for this book from GitHub at `https://github.com/PacktPublishing/Neural-Search-From-Prototype-to-Production-with-Jina`. If there's an update to the code, it will be updated in the GitHub repository.

We also have other code bundles from our rich catalog of books and videos available at `https://github.com/PacktPublishing/`. Check them out!

Download the color images

We also provide a PDF file that has color images of the screenshots and diagrams used in this book. You can download it here: https://packt.link/minUU

Conventions used

There are a number of text conventions used throughout this book.

Code in text: Indicates code words in text, database table names, folder names, filenames, file extensions, pathnames, dummy URLs, user input, and Twitter handles. Here is an example: "To interact with the multi-modal application in the web browser via the UI, you can use the index. html HTML file provided in the static folder."

A block of code is set as follows:

```
- name: keyValueIndexer
  uses:
    jtype: KeyValueIndexer
    metas:
      workspace: ${{ ENV.HW_WORKDIR }}
      py_modules:
        - ${{ ENV.PY_MODULE }}
  needs: segment
- name: joinAll
  needs: [textIndexer, imageIndexer, keyValueIndexer]
```

When we wish to draw your attention to a particular part of a code block, the relevant lines or items are set in bold:

```
- name: keyValueIndexer
  uses:
    jtype: KeyValueIndexer
    metas:
      workspace: ${{ ENV.HW_WORKDIR }}
      py_modules:
        - ${{ ENV.PY_MODULE }}
  needs: segment
- name: joinAll
  needs: [textIndexer, imageIndexer, keyValueIndexer]
```

Any command-line input or output is written as follows:

```
<jina.types.arrays.document.DocumentArray length=3 at
5701440528>

{'id': '6a79982a-b6b0-11eb-8a66-1e008a366d49', 'tags': {'id':
3.0}},
{'id': '6a799744-b6b0-11eb-8a66-1e008a366d49', 'tags': {'id':
2.0}},
{'id': '6a799190-b6b0-11eb-8a66-1e008a366d49', 'tags': {'id':
1.0}}
```

Bold: Indicates a new term, an important word, or words that you see onscreen. For instance, words in menus or dialog boxes appear in **bold**. Here is an example: "Select **System info** from the **Administration** panel."

> **Tips or Important Notes**
> Appear like this.

Get in touch

Feedback from our readers is always welcome.

General feedback: If you have questions about any aspect of this book, email us at customercare@ packtpub.com and mention the book title in the subject of your message.

Errata: Although we have taken every care to ensure the accuracy of our content, mistakes do happen. If you have found a mistake in this book, we would be grateful if you would report this to us. Please visit www.packtpub.com/support/errata and fill in the form.

Piracy: If you come across any illegal copies of our works in any form on the internet, we would be grateful if you would provide us with the location address or website name. Please contact us at copyright@packt.com with a link to the material.

If you are interested in becoming an author: If there is a topic that you have expertise in and you are interested in either writing or contributing to a book, please visit authors.packtpub.com.

Share Your Thoughts

Once you've read *Neural Search - From Prototype to Production with Jina*, we'd love to hear your thoughts! Scan the QR code below to go straight to the Amazon review page for this book and share your feedback.

https://packt.link/r/1801816824

Your review is important to us and the tech community and will help us make sure we're delivering excellent quality content.

Part 1: Introduction to Neural Search Fundamentals

In this part, you will understand what Neural Search is and what it is used for. You'll see an overview of its practical applications, the challenges it poses, and how to overcome them. The following chapters are included in this part:

- *Chapter 1, Neural Networks for Neural Search*
- *Chapter 2, Introducing Foundations of Vector Representation*
- *Chapter 3, System Design and Engineering Challenges*

1

Neural Networks for Neural Search

Search has always been a crucial part of all information systems; getting the right information to the right user is integral. This is because a user query, as in a set of keywords, cannot fully represent a user's information needs. Traditionally, symbolic search has been developed to allow users to perform keyword-based searches. However, such search applications were bound to a text-based search box. With the recent developments in deep learning and artificial intelligence, we can encode any kind of data into vectors and measure the similarities between two vectors. This allows users to create a query with any kind of data and get any kind of search result.

In this chapter, we will review important concepts regarding information retrieval and neural search, as well as looking at the benefits that neural search provides to developers. Before we start introducing neural search, we will first introduce the drawbacks of the traditional symbolic-based search. Then, we'll move on to looking at how to use neural networks in order to build a cross/multi-modality search. This will include looking at its major applications.

In this chapter, we're going to cover the following main topics in particular:

- Legacy search versus neural search
- Machine learning for search
- Practical applications for neural search

Technical requirements

This chapter has the following technical requirements:

- **Hardware**: A desktop or laptop computer with a minimum of 4 GB of RAM; 8 GB is suggested
- **Operating system**: A Unix-like operating system such as macOS, or any Linux-based distribution, such as Ubuntu
- **Programming Language**: Python 3.7 or higher, and Python Package Installer, or pip

Legacy search versus neural search

This section will guide you through the fundamentals of symbolic search systems, the different types of search applications, and their importance. This is followed by a brief description of how the symbolic search system works, with some code written in Python. Last but not least, we'll summarize the pros and cons of the traditional symbolic search versus neural search. This will help us to understand how a neural search can better bridge the gap between a user's intent and the retrieved documents.

Exploring various data types and search scenarios

In today's society, governments, enterprises, and individuals create a huge amount of data by using various platforms every day. We live in the era of big data, where things such as texts, images, videos, and audio files play a significant role in society and the fulfillment of daily tasks.

Generally speaking, there are three types of data:

- **Structured data:** This includes data that is logically expressed and realized by a two-dimensional table structure. Structured data strictly follows a specific data format and length specifications and is mainly stored and managed using relational databases.

- **Unstructured data:** This has neither a regular or complete structure nor a predefined data model. This type of data is not appropriately managed by representing the data using a two-dimensional logical table used in databases. This includes office documents, text, pictures, hypertext markup language (HTML), various reports, images, and audio and video information in all formats.

- **Semi-structured data:** This falls somewhere between structured and unstructured data. It includes log files, **Extensible Markup Language (XML)**, and **Javascript Object Notation (JSON)**. Semi-structured data does not conform to the data model structure associated with relational databases or other data tables, but it contains relevant tags that can be used to separate semantic elements that are used to stratify records and fields.

Search indices are widely used to hunt for unstructured and semi-structured data within a massive data collection to meet the information needs of users. Based on the levels and applications of the document collection, searches can be further divided into three types: web search, enterprise search, and personal search.

In a **web search**, the search engine first needs to index hundreds of millions of documents. The search results are then returned to users in an efficient manner while the system is continuously optimized. Typical examples of web search applications are Google, Bing, and Baidu.

In addition to web search, as a software development engineer, you are likely to encounter *enterprise* and *personal search* operations. In enterprise search scenarios, the search engine indexes internal documents of an enterprise to serve the employees and customers of the business, such as an internal patent search index of a company, or the search index of a music platform, such as SoundCloud.

If you are developing an email application and intend to allow users to search for historical emails, this constitutes a typical example of a personal search. This book focuses on enterprise and personal types of search operations.

Important Note

Make sure you understand the difference between search and match. Search, in most cases, is done in documents organized in an unstructured or semi-structured format, while match (such as an SQL-like query) takes place on structured data, such as tabular data.

As for different data types, the concept of modality plays an important role in a search system. Modality refers to the form of information such as text, images, video, and audio files. Cross-modality search (also known as *cross-media search*) refers to retrieving samples from different modes with similar semantics by exploring the relationship between different modalities and employing a certain modal sample.

For example, when we enter a keyword in an email inbox application, we can find the appropriate email returns as a result of a unimodal search – searching text by text. When you enter a keyword on a page for image retrieval, the search engine will return appropriate images as a result of a cross-modal search, searching images by text.

Of course, a unimodal search is not limited to searching text by text. The app known as Shazam, which is popular in the App Store, helps users to identify music and returns a track's title to users in a short time. This can be seen as an application of unimodal search. Here, the concept of modality no longer refers to text, but to audio. On Pinterest, users can locate similar images through an image search, where the modality refers to an image. Likewise, the scope of a cross-modal search covers far more than searching for images by text.

Let's consider this from another perspective. Is it possible for us to search across multiple modalities? Of course, the answer is "Yes!" Imagine a search scenario where a user uploads a photo of clothes and wants to look for similar types of clothing (we usually call this type of application "shop the look"), and at the same time enters a paragraph that describes the clothes in the search box to improve the accuracy of the search. In this way, our search keywords span two modalities (text and images). We refer to this search scenario as a multi-modal search.

Now that we have a grasp of the concept of modality, we will elaborate on the working principles, advantages, and disadvantages of symbolic search systems. By the end of this section, you will understand, that symbolic search systems cannot deal with different modalities.

How does the traditional search system work?

As a developer, you may have used Elasticsearch or Apache Solr to build a search system in web applications. These two widely used search frameworks were developed based on Apache Lucene. We'll take Lucene as a case in point to introduce the components of a search system. Imagine you intend to search for a keyword in thousands of text documents (`txt`). How will you complete this task?

The easiest solution is to traverse all text documents from a path and read through the contents of these documents. If the keyword is in the file, the name of the document will be returned:

```python
# src/chapter-1/sequential_match.py
import os
import glob


dir_path = os.path.dirname(os.path.realpath(__file__))

def match_sequentially():
    matches = []
    query = 'hello jina'
    txt_files = glob.glob(f'{dir_path}/resources/*.txt')
    for txt_file in txt_files:
        with open(txt_file, 'r') as f:
            if query in f.read():
                matches.append(txt_file)
    return matches


if __name__ == '__main__':
    matches = match_sequentially()
    print(matches)
```

The code fulfills the simplest search function by traversing all files with the extension .txt in the current directory and then opening those files in turn. If the keyword hello jina used for the query is available, the filename will be printed with all the matching files. Although these lines of code allow you to conduct a basic search, the process has many flaws:

- **Poor scalability**: In a production environment, there may be millions of files to be retrieved. Meanwhile, users of the retrieval system expect to obtain retrieval results in the shortest possible time, posing stringent requirements for the performance of the search system.

- **Lack of a relevance measurement**: The code helps you achieve the most basic Boolean retrieval, which is to return the result of a match or mismatch. In a real-world scenario, users need a score to measure the relevance degree from a search system that is sorted in descending order, with more relevant files being returned to users first. Obviously, the aforementioned code snippets are unable to fulfill this function.

To address these issues, we need to *index* the files to be retrieved. **Indexing** refers to a process of converting a file type that allows a rapid search and skipping the continuous scanning of all files.

As an important part of our daily lives, indexing is comparable to consulting a dictionary and visiting a library. We'll use the most widely used search library, Lucene, to illustrate the idea.

Lucene Core (`https://lucene.apache.org/`) is a Java library providing powerful indexing and search features, as well as spellchecking, hit highlighting, and advanced analysis/tokenization capabilities. Apache Lucene sets the standard for search and indexing performance. It is the search core of both Apache Solr and Elasticsearch.

In Lucene, after all collections of files to be retrieved are loaded, you may extract texts from such files and convert them to Lucene Documents, which generally contain the title, body, abstract, author, and URL of a file.

Next, your file will be analyzed by Lucene's *text analyzer*, which generally includes the following processes:

Tokenizer: This splits the raw input paragraphs into tokens that cannot be further decomposed.

Decomposing compound words: In languages such as German, words composed of two or more tokens are called compound words.

Spell correction: Lucene allows users to conduct spellchecking to enhance the accuracy of retrieval.

Synonym analysis: This enables users to manually add synonyms in Lucene to improve the recall rate of the search system (note: the accuracy rate and recall rate will be elaborated upon shortly).

Stemming and lemmatization: The former enables users to derive the root by removing the suffix of a word (for example, *play*, the root form, is derived from the words *plays*, *playing*, and *played*), while the latter helps users convert words into basic forms, such as *is*, *are*, and *been*, which are converted to *be*.

Let's attempt to preprocess some texts using **NLTK**.

> **Important Note**
> NLTK is a leading platform for building Python programs to work with human language data. It provides easy-to-use interfaces to over 50 corpora and lexical resources.

First, install a Python package called `nltk` with this command:

```
pip install nltk
python -m nltk.downloader 'punkt'
```

We preprocess the text `Jina is a neural search framework built with cutting-edge technology called deep learning`:

```
import nltk

sentence = 'Jina is a neural search framework built with
cutting-edge technology called deep learning'

def tokenize_and_stem():
    tokens = nltk.word_tokenize(sentence)
    stemmer = nltk.stem.porter.PorterStemmer()
    stemmed_tokens = [stemmer.stem(token) for token in
                        tokens]
    return stemmed_tokens


if __name__ == '__main__':
    tokens = tokenize_and_stem()
    print(tokens)
```

This code enables us to carry out two operations on a sentence: tokenizing and stemming. The results of each are printed respectively. The raw input strings are parsed into a list of strings in Python, and finally each parsed token is lemmatized to its basic form. For instance, `cutting` and `called` are respectively converted to `cut` and `call`. For more operations, please refer to the official documentation of NLTK (`https://www.nltk.org/`).

After files are processed with the Lucene Document, the *clean* files will be indexed. Generally, in a traditional search system, all files are indexed using an **inverted index**. An inverted index (also referred to as a **postings file** or **inverted file**) is an index data structure storing a map of content, such as words or numbers, to its locations in a database file, or in a document or a set of documents.

Simply put, an inverted index consists of two parts: a **term dictionary**, and **postings**.

Tokens, their IDs, and the document frequency (the frequency of such tokens appearing in the entire collection of documents to be retrieved) are stored in the term dictionary. A collection of all tokens is called a vocabulary. All tokens are sorted in alphabetical order in the dictionary.

In the postings, we save the token ID and the document IDs where the token occurred. Assuming that in the aforesaid example, the token `jina` from our query keyword `hello jina` appears three times in the entire collection of documents (in `1.txt`, `3.txt`, and `11.txt`), then the token is "jina" and the document frequency is 3. Meanwhile, the names of the three text documents, `1.txt`, `3.txt`, and `11.txt`, are saved in the posting. Then, the indexing of the text file is completed as shown in the following figure:

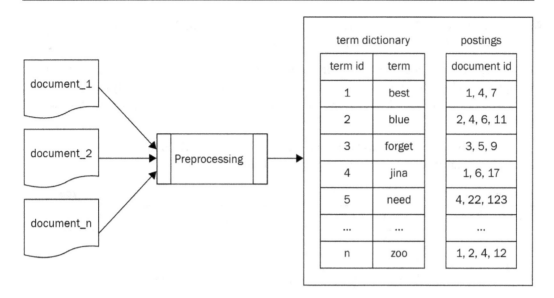

Figure 1.1 – Data structure of inverted index

When a user makes a query, keywords used for the query are generally shorter than the collection of documents to be retrieved. Lucene can perform the same preprocessing for such keywords (such as tokenization, decomposition, and spelling correction).

The processed tokens are mapped to the postings through the term dictionary in the inverted index so that matched files can be quickly found. Finally, Lucene's scoring starts to work and scores each related file discovered according to a vector space model. Our index file is stored in an inverted index, which may be represented as a vector.

Assuming that our query keyword is jina, we map it to the vector of the inverted index and have it represented by - when it does not appear in the file; then the query vector [-, 'jina', -, -, ...] can be obtained. This is how we represent a *query*, as a **vector space model**, in a traditional search engine.

vocabulary	best	blue	forget	jina	need	...	zoo
query	-	-	-	True	-	-	-
d1	True	-	-	True	-	-	True
d2	-	True	-	-	-	-	True
...	-	-	-	-	-	-	-
dn	-	-	-	-	-	-	True

Figure 1.2 – Term occurrence in the vector space model

Next, in order to derive the ranking, we need to numerically represent the token of the space vector model. Generally, **tf-idf** is regarded as a simple approach.

With this algorithm, we grant a higher weight to any token that appears relatively frequently. If such a token appears multiple times in many documents, we believe that the token is weakly representative, and the weight of the token will be reduced again. If the token does not appear in the documents, its weight is 0.

In Lucene, an algorithm called **bm-25** is employed more frequently, which further optimizes tf-idf. After numerical calculation, the vector is expressed as follows:

vocabulary	best	blue	forget	jina	need	...	zoo
query	0	0	0	1	0	0	-
d1	0.0268	0	0	0.124	0	0	0.0155
d2	0	0.0335	0	0	0	0	1
...	0	0	0	0	0	0	0
dn	0	0	0	0	0	0	0.0155

Figure 1.3 – Vector space representation

As shown in the preceding figure, because the word **a** appears too frequently, it appears in document 1 and document 2 and has a low weight score. The token *jina*, a relatively uncommon word (appearing in document 2), has been granted a higher weight.

In the query vector, because the query keyword only has one word, *jina*, its weight is set as 1 and the weights of other tokens that do not appear are set as 0. Afterward, we multiply the query vector and the document vector element by element and add up the results to obtain the score of each document corresponding to the query keyword. Then, reverse sorting is performed so that the sorted documents can be returned to the user according to the score sorted in an inverted order (from high to low scores).

In short, if the keyword used for a query appears more frequently in a particular file and less frequently in the vocabulary file, its relative score will be higher and returned to the user with a higher priority. Of course, Lucene also grants different weights to various parts of a file. For example, the title and keywords of the file will have a higher scoring weight than the body would. Given the fact this book is about neural search, this aspect will not be elaborated upon further here.

Pros and cons of the traditional search system

In the previous section, we briefly revisited traditional symbolic search. Perhaps you have noticed that both the Lucene we introduced previously and the Lucene-based search frameworks, such as

Elasticsearch and Solr, are based on text retrieval. This has quite a few advantages in the application scenarios of searching text by text:

Mature technology: Since research and development were done in 1999, the Lucene and Lucene-based search systems have existed for over 20 years and have been widely used in various web applications.

Easy integration: As users, developers of a web application do not need to have a deep understanding of Elasticsearch, Solr, or the operating logic of Lucene; only a small amount of code is required to integrate a high-performing, extensible search system into web applications.

Well-developed ecosystem: Thanks to the operation of Elastic Company, Elasticsearch has extended its search system functionality significantly. Currently, it is not only a search framework, but also a platform equipped with user management, a restful interface, data backup and restoration, and security management such as single sign-in, log audit, and other functions. Meanwhile, the Elasticsearch community has contributed a variety of plugins and integrations.

At the same time, you have probably realized that both Elasticsearch and Solr with Lucene at the core have unavoidable flaws.

In the previous section, we introduced the concept of modality. Lucene and Elasticsearch, which is built on top of it, are inherently unable to support cross-modal and multi-modal search options. Let's take a moment to review the operating principle of Lucene, as Lucene has powered most of the search systems users are using on a daily basis. When texts are preprocessed in the first place, the search keyword must be text. When a data collection to be retrieved is preprocessed and indexed, likewise the index result is also the text stored in the inverted index.

In this way, the Lucene-based search platform can only rely on the text modality and retrieve data in the text modality. If objects to be retrieved are images, audio, or video files, how can they be found using a traditional search system? It is quite simple; two main methods are employed:

Manual tagging and adding metadata: For example, when a user uploads a song to a music platform, they may manually tag the author, album, music type, release time, and other data. Doing so ensures that users are able to retrieve music using text.

Hypothesis of the surrounding text: If an image, in the absence of user tagging, appears in an article, it will be assumed by the traditional search system to be more closely associated with its surrounding text. Accordingly, when a user's query keyword matches the surrounding text of the image, the latter will be matched.

The essence of the two methods is to convert the document of non-text modality into a text modality so as to effectively use the current retrieval technology. However, this modal conversion process either relies on a large amount of manual tagging, or is done at the cost of query accuracy, which greatly undermines the user's search experience.

Likewise, this type of search mode limits the user's search habits to a keyword search and cannot be extended to a real cross-modal or even multi-modal search. For deeper insight into this issue, we may use a vector space to represent keywords of a paragraph and use another vector space to denote a text document to be retrieved. However, due to the restrictions of the technology back in the days when we had to rely on traditional search systems, we were unable to use the space vector to represent a piece of music, image, or video. It is also impossible to map two documents of different modalities to the same space vector to compare their similarities.

With the research and development on (statistical) machine learning techniques, more and more researchers and engineers have started to empower their search system using machine learning algorithms.

Machine learning for search

As a cross-disciplinary task, neural search has gone beyond the boundaries of information retrieval. It requires a general understanding of the concepts of machine learning, deep learning, and how we can apply these techniques to improve a search task. In this section, we will give a brief introduction to machine learning and how it can be applied to search systems.

Understanding machine learning and artificial intelligence

Machine learning refers to a technique that teaches computers to make decisions in a way that comes naturally to humans by enabling computers to learn the inherent laws of data and acquire new experience and knowledge, thus improving their intelligence.

Because various industries require an increased level of efficiency during data processing and analysis due to their growing demand for data, a large number of machine learning algorithms have emerged. The concept of statistical machine learning algorithms primarily refers to the steps and processes of solving optimization problems through mathematical and statistical methods.

With respect to different data and model requirements, appropriate machine learning algorithms are selected and employed to tackle practical issues in a more efficient manner. Machine learning has achieved great success in many fields, such as natural language understanding, computer vision, machine translation, and expert systems. It is fair to say that whether a system has a *learning* function or not has become a hallmark of it possessing intelligence.

Hinton et al. (2006) proposed the concept of **deep learning** (deep learning/deep neural networks). In 2009, Hinton introduced deep neural networks to scholars specialized in voices. Hence, in 2010, this field of research witnessed a remarkable breakthrough in speech recognition. In the next 11 years, **convolutional neural networks** (CNNs) were applied in the field of image recognition, leading to significant achievements.

Three founders of neural networks, LeCun, Bengio, and Hinton (2015), published a review titled *Deep Learning* in Nature. This shows that deep neural networks have not only been accepted by academia, but also the industrial field. Furthermore, in 2016 and 2017, the world witnessed a general expansion in deep learning. AlphaGo and AlphaZero were invented by Google after a short learning period and won a landslide victory over the top three Go players in the world. The intelligent voice system launched by iFLYTEK boasts a recognition accuracy rate of over 97% and stands at the forefront of AI worldwide; the autonomous driving systems developed by companies such as Google and Tesla have passed a milestone of testing on the road. These achievements have unveiled the value and charm of neural networks to humans again.

Machine learning has been applied to various industries, so maybe we can ask ourselves: can we apply machine learning to search applications? The answer here is "Yes." In the next section, we'll give a brief overview of different types of machine learning and how search can benefit from it.

Machine learning and learning-to-rank

Imagine a scenario where you intend to train a model capable of evaluating the price of a new apartment or house based on the collected data related to local real estate information and prices. This is one of the most important tasks of machine learning: **regression**.

Before the popularization of the deep learning technique, data analysts would have had to clean this data, use business logic to perform feature engineering, and design features of a real estate price predictor, such as the floor area, construction time, and type of apartments or houses, as well as the average prices of surrounding apartments or houses, and so on.

After feature engineering has been completed, raw data will be used to form a two-dimensional data table similar to Excel. The horizontal axis represents each house record, and the vertical axis represents each feature. The data is usually divided into two to three parts again: the majority of the data is used for *model training*, while a small amount of the data is used for *model evaluation*.

Next, machine learning engineers will select one or more appropriate algorithms from the machine learning toolkit to train the model and evaluate the performance of the model in the test data. Finally, the model with the best performance will be deployed in the production environment to serve customers.

Imagine another scenario where many landmark pictures have been collected from social networks. When a user uploads a new landmark picture, you expect your system to automatically recognize the name of the site. This is another important task of machine learning: **classification**.

In the field of traditional machine learning and computer vision, some features, such as SIFT, SURF, and HOG, are employed to develop a **Bag-of-Visual-Words (BoW)** through which a vector representation of this photo is established. Moreover, models are used to predict the classification. Nowadays, deep learning serves as the model to extract visual features from images without the requirement of feature engineering.

Let's take a moment to look at our two examples. During the training process of predicting prices of houses (apartments), models are trained using *feature engineering*. All the training data is ground-truth, i.e., the house (apartment) prices and landmark names are documented. Such tasks are collectively referred to as supervised machine learning in the field of machine learning.

Since we can perform regression analysis and classification of data through supervised machine learning, is it possible to apply supervised machine learning in the search? The answer is yes, of course.

Assuming that our task is to optimize the search system, the goal is to predict the user's click rate for the document and return documents with a higher predicted click rate to the user first. This is called **learning-to-rank** (first stage) and **neural information retrieval** (second stage). The concept of learning-to-rank (based on statistical machine learning), as proposed by academia in the early 1990s, evolved for nearly 20 years before experiencing a downturn since the emergence of deep learning in 2010, when neural information retrieval was at its peak.

Just like the prediction of an apartment (or house) price, or landmark recognition, engineers first performed data engineering after collecting such data. Common features include the number of query keywords in the document title/body, the percentage of the document title body that contains the query keywords, the tf-idf score, and the bm-25 score, among others. It follows that the final score of a traditional search system is used as a numerical feature during the training of models.

In a real-world scenario, Microsoft's Bing search platform was designed with its own *Microsoft Learning to Rank Datasets*, which contain 136 features. Besides, they published a contest for learning-to-rank, calling for the use of these datasets as a basic training model for predicting the matching degree of web pages. After that, a trained model is applied in the production environment of a Bing search, which has improved the search effect to a certain extent.

At the same time, search companies such as Google, Yahoo, and Baidu have also conducted a large amount of research and have partially deployed their research results into the production environment.

In the fields of enterprise and personal search, Elastic has developed its learning-to-rank plugin named ElasticSearch LTR, which can be plug-in into your ES-powered search system. As a user, you still need to use a familiar machine learning framework to design features, train the learning-to-rank model, evaluate the model performance, and select models. Elasticsearch's support for learning-to-rank can be plugged into the existence search system and get a new predicted ranking score based on model output. Although machine learning can be used to design models for multi-modal data, Elasticsearch places more emphasis on text-to-text search. Figure 1.4 demonstrates how learning-to-rank works in a search system.

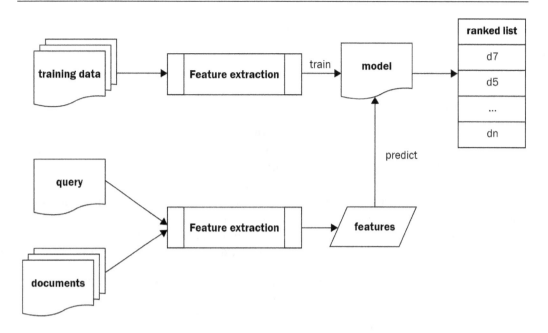

Figure 1.4 – Learning-to-rank

This book will focus on search powered by deep neural networks, namely **neural information retrieval**.

The advantage of neural information retrieval is that users do not have to *design* features by themselves. Normally, we leverage two independent deep learning models (neural nets) as feature extractors to extract vectors from queries and documents respectively. Then, we measure the similarity between two vectors using metrics, such as cosine similarity. At this stage, neural-network-powered search has become very promising for industrial use cases. In the next section, we will introduce some of the potential applications for neural-network-powered search.

Practical applications powered by neural search

The previous section provided an overview of the representation and principles of dense vectors. This section will focus on the application of these vectors. During our daily work and study, all files will have a unique modality, such as a text, image, audio, or video file, and so on. If documents of any modality can be represented by dense vectors and then mapped to the same vector space, it is possible to compare the cross-modal similarity. This also allows us to use one modality to search for data in another modality.

This scenario was first extensively put into practice in the field of e-commerce with the common use of *image search*, for example. Its major application in this field includes having a product photo and hunting for related or similar products offline and online.

The e-commerce search primarily consists of steps such as the following:

1. **Preprocessing**
2. **Feature extraction and fusion**
3. **Large-scale similarity search**

During preprocessing, techniques such as **resizing**, **normalization**, and **semantic segmentation** may first be employed to process images. Resizing and normalization enable the input image to match the input format of the pre-trained neural network. Semantic segmentation has the function of removing background noise from the image and leaving only the product itself. Of course, we need to pre-train a neural pathway for feature extraction, which will be elaborated on shortly. By the same token, if the dataset of an e-commerce product to be retrieved has a large amount of noise, such as a large number of buildings, pedestrians, and so on in the background of fashion photos, it will be necessary to train a semantic segmentation model that can help us accurately extract the product profile from photos.

During feature extraction, a **fully connected** (FC) layer of deep learning is generally used as a feature extractor. The common backbone models of deep learning are AlexNet, VGGNet, Inception, and ResNet. These models are usually pre-trained on a large-scale dataset (such as the ImageNet dataset) to complete classification tasks. Transfer learning is carried out with the dataset in the e-commerce field in a bid to make the feature extractor suitable for the field, such as the feature extraction of fashion. Currently, a feature extractor with deep learning techniques at its core can be regarded as a *global feature extractor*. In some applications, traditional computer vision features, such as SIFT or VLAD, are employed for the extraction of local features and fusion with global features to enhance vector representation. The global feature will transform the preprocessed image into a dense vector representation.

When users make a query based on the search for images with images, the keyword used for the query is also an image. The system will generate a dense vector representation of that image. Then, users will be able to find the most similar image by comparing the dense vector of the image to be queried against those of all images in the library. This is feasible in theory. However, in reality, with the rapid increase in the number of commodities, there may be tens of millions of dense vectors of indexed images. As a result, the comparison of vectors in a pair-wise manner will fail to meet the user's requirements for a quick response from the retrieval system.

Therefore, *large-scale similarity search techniques*, such as product quantization, are generally used to divide the vector to be searched into multiple buckets and perform a quick match based on the buckets by minimizing the recall rate and greatly speeding up the vector-matching process. Therefore, this technique is also commonly referred to as *approximate nearest neighbor*, or *ANN retrieval*. Commonly used ANN libraries include the FAISS, which is maintained by Facebook, and Annoy, maintained by Spotify.

Likewise, the search for images by images in an e-commerce scenario is also applicable to other scenarios, such as *Tourism Landmark Retrieval* (using pictures of tourist attractions to quickly locate other pictures of that attraction or similar tourist attractions), or *Celebrity Retrieval* (used to find photos of celebrities and retrieve their pictures). In the field of search engines, there are many such applications, which are collectively referred to as *reverse image search*.

Another interesting application is **question answering**. Neural-network-based search systems could be powerful when building a question-answering (QA) system on different tasks. First, the questions and answers that are currently available are taken as a training dataset on which to develop a pre-trained model of texts. When the user enters a question, the pre-trained model is employed to encode the question into a dense vector representation, conduct similarity matching in the dense vector representation of the existing repository of answers, and quickly help users find the answer to a question. Second, many question-answering systems, such as Quora, StackOverflow, and Zhihu, already have a large number of previously asked questions. When a user wants to ask a question, the question-answering system first determines whether the question has already been asked by someone else. If so, the user will be advised to click and check the answers to similar questions instead of repeating the query. This also involves similarity match, which is normally referred to as *deduplication* or *paraphrase identification*.

Meanwhile, in the real world, a large number of unexplored applications can be completed using neural information retrieval. For instance, if you employ text to search for untagged music, it is necessary to map the text and music representation to the same vector space. Then, the appearance time of scenarios in the video can be located using images. Conversely, when a user is watching a video, a product that appears in the video is retrieved and the purchase can be completed. Alternatively, deep learning can be carried out for specialized data retrieval, such as source code retrieval, DNA sequence retrieval, and more!

New terms learned in this chapter

- **Traditional search**: Mostly applied to text retrieval. Measures the similarity by the weighted score of occurrences of a set of tokens from a query and documents.

- **Indexing**: The process of converting files that allow a rapid search and skipping the continuous scanning of all files.

- **Searching**: The process of conducting similarity score computation against a user query and indexed documents inside the document store and returning the top-k matches.

- **Vector space model**: A way to represent a document numerically. The dimension of the VSM is the number of distinct tokens in all documents. The value of each dimension is the weight of each term.

- **TF-IDF**: Term-Frequency Inverse Document Frequency is an algorithm that is intended to reflect how important a word is to a document in a collection of documents that are to be indexed.

- **Machine learning**: This refers to a technique that teaches computers to make decisions in a way that comes naturally to humans by enabling computers to learn the distribution of data and acquire new experience and knowledge.

- **Deep neural networks**: A **deep neural network** (**DNN**) is an **artificial neural network** (**ANN**) with multiple layers between the input and output layers that aims to predict, classify, or learn a compact representation (dense vector) of a piece of data.

- **Neural search**: Unlike symbolic search, neural search makes use of the representation (a dense vector) generated by DNNs and measures the similarity between a query vector and a document vector, returning the top-k matches based on certain metrics.

Summary

In this chapter, you have learned about the key concepts of searching and matching. We have also covered the difference between legacy search and neural-network-based search. We saw how neural networks can help us tackle the issues traditional search cannot solve, such as cross-modality or multi-modality search.

Neural networks are able to encode different types of information into a common embedding space and make different pieces of information comparable, and that's why deep learning and neural networks have the potential to better fulfill a user's information needs.

We have introduced several possible applications using deep-learning-powered search systems, for instance, vision-based product search in fashion or tourism, or text-based search for question answering and text deduplication. More kinds of application are still to be explored!

You should now understand the core idea behind neural search: neural search has the ability to encode any kind of data into an expressive representation, namely an **embedding**. Creating a quality embedding is crucial to a search application powered by deep learning, since it determines the quality of the final search result.

In the next chapter, we will introduce the foundations of embeddings, such as how to encode information into embeddings, how to measure the distance between different embeddings, and some of the most important models we can use to encode different modalities of data.

2
Introducing Foundations of Vector Representation

Vectors and **vector representation** are at the very core of neural search since the quality of vectors determines the quality of search results. In this chapter, you will learn about the concept of **vectors** within **machine learning** (**ML**). You will see common search algorithms using vector representation as well as their weaknesses and strengths.

We're going to cover the following main topics in this chapter:

- Introducing vectors in ML

- Measuring the similarity between two vectors

- Local and distributed representations

By the end of this chapter, you will have a solid understanding of how every type of data can be represented in vectors and why this concept is at the very core of neural search.

Technical requirements

This chapter has the following technical requirements:

- A laptop with a minimum of 4 GB RAM (8 GB or more is preferred)

- Python installed with version 3.7, 3.8, or 3.9 on a Unix-like operating system, such as macOS or Ubuntu

The code for this chapter can be found at `https://github.com/PacktPublishing/Neural-Search-From-Prototype-to-Production-with-Jina/tree/main/src/Chapter02`.

Introducing vectors in ML

Text is an important means of recording human knowledge. As of June 2021, the number of web pages indexed by mainstream search engines such as Google and Bing has reached 2.4 billion, and the majority of information is stored as text. How to store this textual information, and even how to efficiently retrieve the required information from the repository, has become a major issue in information retrieval. The first step in solving these problems lies in representing text in a format that is *comprehensible* to computers.

As network-based information has become increasingly diverse, in addition to text, web pages contain a large amount of multimedia information, such as pictures, music, and video files. These files are more diverse than text in terms of form and content and satisfy users' needs from different perspectives. How to represent and retrieve these types of information, as well as how to pinpoint the multimodal information needed by users from the vast mass of data available on the internet is also an important factor to be considered in the design of search engines. To achieve this, we need to represent each document as its vector representation.

A *vector* is an object that has both a magnitude and a direction, as you may remember learning in school. If we can represent our data using vector representation, then we're able to use the angle to measure the similarity of two pieces of information. To be more concrete, we can say the following:

- Two pieces of information are represented as vectors
- Both vectors start from the origin [0, 0] (assuming two dimensions)
- Two vectors form an angle

Figure 2.1 illustrates the relationship between two vectors with respect to their angle:

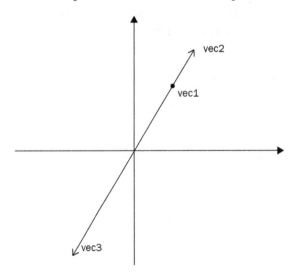

Figure 2.1 – An example of vector representation

vec1 and **vec2** have the same direction but different lengths. **vec2** and **vec3** have the same lengths but point in opposite directions. If the angle is 0 degrees, the two vectors are identical. If the vector is 180 degrees, the two vectors are completely opposite. We can measure the similarity between two vectors by the angle: the smaller the angle, the closer the vectors are. This method is also called **cosine similarity**.

In reality, cosine similarity is one of the most commonly used similarity measurements to determine the similarity between two vectors, but not the only one. We'll dive into it in more detail, as well as other similarity metrics, in the *Measuring the similarity between two vectors* section. Before that, you might be wondering how we can encode our raw information, such as text or audio, into a vector of numeric values. In this section, we're going to do that.

We'll dive into the details of cosine similarity using *Python* and the *NumPy library*. As well as that, we will introduce other similarity metrics and briefly cover local and distributed vector representation in the following subsections.

Using vectors to represent data

Let's start with the most common scenario: **representing text information**.

First of all, let's define the concept of a **feature vector**. Let's say we want to build a search system for Wikipedia (in English). As of July 2022, English Wikipedia has over 6.5 million articles containing over 4 billion words (180,000 unique words). We can call these unique words the Vocabulary of Wikipedia.

Each of the articles in this Wikipedia collection should be encoded into a series of numerical values; this is referred to as a feature vector. To this end, we can encode 6.5 million articles into 6.5 million indexed feature vectors, then use a similarity metric, such as cosine similarity, to measure the similarity between the encoded query feature vector and the indexed 6.5 million feature vectors.

The encoding process involves finding an optimal function to transform the original data into its vector representation. How can we achieve this goal?

Again, we start with the simplest method: using a **bit vector**. A bit vector means all the values inside the vector will be either 0 or 1, depending on the occurrence of the word. Let's say we loop over all unique words in the Vocabulary; if the word occurs in this particular document, *d*, then we set the value of the location of this unique word to be 1, otherwise 0.

Let's refresh what we introduced in *Chapter 1, Neural Networks for Neural Search*, in the *How does the traditional search system work?* section, imagining we have two documents:

- doc1 = *Jina is a neural search framework*

- doc2 = *Jina is built with cutting edge technology called deep learning*

1. If we merge these two documents, we have a Vocabulary (of unique words), as follows:

    ```
    vocab = 'Jina is a neural search framework built with
    cutting age technology called deep learning'
    ```

2. Imagining the preceding variable, `vocab`, is our Vocabulary, after preprocessing (tokenizing and stemming), we get a list of tokens, as follows:

```
vocab = ['a', 'age', 'built', 'call', 'cut', 'deep',
'framework', 'is', 'jina', 'learn', 'neural', 'search',
'technolog', 'with']
```

Note that the aforementioned Vocabulary has been sorted alphabetically.

3. To encode `doc1` into a vector representation, we loop through all the words inside the **Vocabulary**, check the occurrence of the word within `doc1`, and create the bit vector:

```
import nltk


doc1 = 'Jina is a neural search framework'
doc2 = 'Jina is built with cutting age technology called
deep learning'



def tokenize_and_stem(doc1, doc2):
    tokens = nltk.word_tokenize(doc1 + doc2)
    stemmer = nltk.stem.porter.PorterStemmer()
    stemmed_tokens = [stemmer.stem(token) for token in
                      tokens]
    return sorted(stemmed_tokens)



def encode(vocab, doc):
    encoded = [0] * len(vocab)
    for idx, token in enumerate(vocab):
        if token in doc:
            encoded[idx] = 1  # token present in doc
    return encoded

if __name__ == '__main__':
    tokens = tokenize_and_stem(doc1, doc2)
    encoded_doc1 = encode(vocab=tokens, doc=doc1)
    print(encoded_doc1)
```

The preceding code block encodes `doc1` into a bit vector. In the `encode` function, we firstly created a Python list filled with 0s; the length of the list is identical to the size of Vocabulary. Then, we loop over the Vocabulary to check the occurrence of the word inside the document to encode. If present, we set the value of the encoded vector as `1`. In the end, we get this:

```
>>> [1, 0, 0, 0, 0, 0, 0, 1, 1, 0, 0, 1, 1, 0, 0]
```

In this way, we've successfully encoded a document into its bit vector representation.

> **Important Note**
>
> You might have noticed that in the preceding example, the output of the bit vector contains a lot of 0s values. In a real-world scenario, as the size of the Vocabulary gets much larger, and the dimensionality of the vector gets very high, there is a high chance that most of the dimensions in the encoded documents are filled with 0s, which is extremely inefficient to store and retrieve. This is also called a **sparse vector**. Some Python libraries, such as SciPy, have strong sparse vector support. Some deep learning libraries, such as TensorFlow and PyTorch, have built-in sparse tensor support. Meanwhile, Jina primitive data types support SciPy, TensorFlow, and PyTorch sparse representations.

So far, we have learned that a vector is an object that has both a magnitude and a direction. We also managed to create the simplest form of vector representation of two text documents using a bit vector. Now, it would be very interesting to know how similar these two documents are. Let us learn more about this in the next section.

Measuring similarity between two vectors

Measuring similarity between two vectors is important in a neural search system. Once all of the documents have been indexed into their vector representation, given a user query, we carry out the same encoding process to the query. In the end, we compare the encoded query vector against all the encoded document vectors to find out what the most similar documents are.

We can continue our example from the previous section, trying to measure the similarity between `doc1` and `doc2`. First of all, we need to run the script two times to encode both `doc1` and `doc2`:

```
doc1 = 'Jina is a neural search framework'
doc2 = 'Jina is built with cutting age technology called deep
learning'
```

Then, we can produce a vector representation for both of them:

```
encoded_doc1 = [1, 0, 0, 0, 0, 0, 0, 1, 1, 0, 0, 1, 1, 0, 0]
encoded_doc2 = [1, 1, 1, 1, 1, 1, 0, 1, 1, 0, 1, 0, 0, 1, 1]
```

Since the dimension of the encoded result is always identical to the size of Vocabulary, the problem has been converted to how to measure the similarity between two vector representations: `encoded_doc1` and `encoded_doc2`.

> **Important Note**
>
> The aforementioned vector representation of `encoded_doc1` and `encoded_doc2` has a depth of 15. It is easy for us to visualize 1D data as a point, 2D data as a line, or 3D data, but not for high-dimensional data. Practically, we might perform dimensionality reduction to reduce high-dimensional vectors to 3D or 2D in order to plot them. The most common technique is called **t-sne**.

Imaging two encoded vector representations can be plotted in a 2D vector space. We can visualize `encoded_doc1` and `encoded_doc2` as follows:

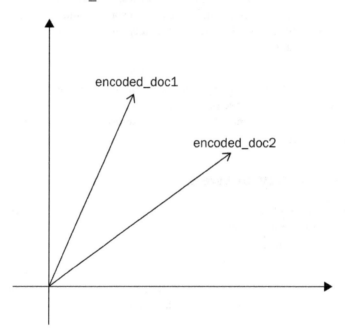

Figure 2.2 – Cosine similarity

Then, we can measure the similarity between `encoded_doc1` and `encoded_doc2` using their angles, specifically, the cosine similarity. The law of cosine tells us that:

$$sim(vec_1, vec_2) = \frac{vec_1 * vec_2}{||vec_1|| * ||vec_2||}$$

Let's say p is represented as [x1, y1] and q is represented as [x2, y2]; then, the aforementioned formula can be rewritten as:

$$sim(vec_1, vec_2) = \frac{(x_1 * y_1) + (x_2 * y_2)}{\sqrt{x_1^2 + x_2^2} + \sqrt{y_1^2 + y_2^2}}$$

Since cosine similarity also works for high-dimensional data, the aforementioned formula can be again rewritten as:

$$sim(vec_1, vec_2) = \frac{\sum_{i=1}^{n}(x_i * y_i)}{\sqrt{\sum_{i=1}^{n} x_i^2} + \sqrt{\sum_{i=1}^{n} y_i^2}}$$

Based on the formula, we can compute the cosine similarity between encoded_doc1 and encoded_doc2, as follows:

```
import math

def compute_cosine_sim(encoded_doc1, encoded_doc2):
    numerator = sum([i * j for i, j in zip(encoded_doc1,
                encoded_doc2)])
    denominator_1 = math.sqrt(sum([i * i for i in
                    encoded_doc1]))
    denominator_2 = math.sqrt(sum([i * i for i in
                    encoded_doc2]))
    return numerator/(denominator_1 * denominator_2)
```

If we print out the result of the similarity between encoded_doc1 and encoded_doc2, we get the following:

```
>>> 0.40451991747794525
```

Here, we get the cosine similarity between two encoded vectors, roughly equal to *0.405*. In a search system, when the user submits a query, we will encode the query into its vector representation. We have encoded all the documents (that we want to search) into their vector representations individually offline. In this way, we can compute the similarity score of the query vector against all document vectors to produce the final ranking list.

> **Important Note**
>
> The preceding code illustrates how you can compute the cosine similarity. The code is not optimized. In reality, you should always use NumPy to perform vectorized operations over vectors (NumPy arrays) to achieve higher performance.

Metrics beyond cosine similarity

Through cosine similarity is the most commonly used similarity/distance metric, there are some other commonly used metrics as well. We will cover another two commonly used distance functions in this section, namely, **Euclidean distance** and **Manhattan distance**.

> **Important Note**
>
> Similarity metrics measure how alike two documents are. On the other hand, distance metrics measure the dissimilarity between two documents. In the search scenario, you always want to get the top k matches against your query. So, if you are using similarity metrics, always get the first k items from the ranked list. On the other hand, while using distance metrics, always get the last k items from the ranked list or reverse the ranked list and get the first k items.

Unlike cosine similarity, which takes the angle of two vectors as its similarity measure, the Euclidean distance takes the length of the line segment between two data points. For instance, consider two 2D docs in the following figure:

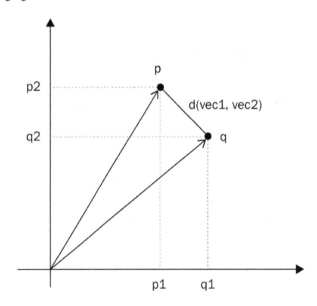

Figure 2.3 – Euclidean distance

As you can see in *Figure 2.3*, previously, we used the angle between `vec1` and `vec2` to compute their cosine similarity. For Euclidean distance, we compute it in a different way. Both `vec1` and `vec2` have a starting point of 0 and the p and q endpoints, respectively. Now, the distance between these two vectors becomes:

$$d(vec_1, vec_2) = \sqrt{(q_1 - p_1)^2 + (q_2 - p_2)^2}$$

Another distance metric is called **Manhattan distance** (or **city-block distance**). It is the distance between two points measured along axes at right angles. In a plane with p at (p1, p2) and q at (q1, q2), the distance between these two vectors becomes:

$$d(vec_1, vec_2) = |p_1 - q_1| + |p_2 - q_2|$$

As can be seen in *Figure 2.4*, the hyperplane has been split into small blocks. Each block has a width of 1 and a height of 1. The distance between p and q becomes 4:

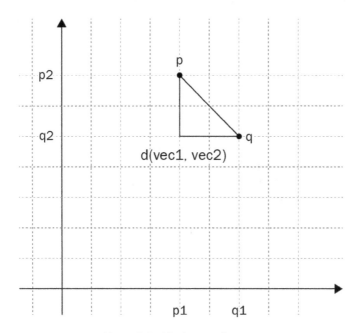

Figure 2.4 – Manhattan distance

There are many other distance metrics as well, such as the **Hamming distance** and **angular distance**, but we won't cover each of them given the fact that cosine and Euclidean are the most commonly used similarity metrics. This, in turn, leads to an interesting question: which distance/similarity metric should I use to make vector similarity computation more effective? The answer is *it depends*.

First of all, it depends on your task and your data. But, in general, when performing text retrieval and related tasks, cosine similarity will be your first choice. It has been widely adopted for applications such as measuring similarity between two pieces of encoded text documents.

The deep learning model might also impact your similarity/distance metric choice. For instance, if you applied metric learning techniques to fine-tune your ML model to optimize certain similarity metrics, then you might stick with the same similarity metric that you optimized. To be more specific, note the following:

- You can apply *Siamese neural networks* to optimize pairs of inputs (query and document) based on the Euclidean distance and get a new model

- When extracting features with the model, it's better to use the *Euclidean distance* as the similarity measure

- If your vectors have extremely high dimensions, it might be a good idea to switch from the Euclidean distance to the *Manhattan distance* since it delivers more robust results

> **Important Note**
>
> In application, different ANN libraries might use different distance metrics as default configuration. For instance, Annoy encourages users to use the angular distance to compute vector distances. It is a variation of the Euclidean distance. More about ANN will be introduced in *Chapter 3, System Design and Engineering Challenges*.

There are multiple ways to encode data into vector representations. Generally speaking, this can be classified into two forms: **local representation** and **distributed representation**. The aforementioned way of encoding data into vector representation can be classified into local representation since it treats each unique word as one dimension.

In the next section, we'll introduce the most important local representation and distributed representation algorithms.

Local and distributed representations

In this section, we'll dive into **local representations** and **distributed representations**. We will go through the characteristics of two different representations and list the most widely used local and global representations to encode different modalities of data.

Local vector representation

As a classic method of text representation, **local representation** only makes use of the **disjointed dimensions** in the vector for a certain word when it is represented as a vector. Disjointed dimension means that each dimensionality of the vector represents a single token.

When only one dimension is used, it is called **one-hot representation**. *One-hot* means that the word is represented as a long vector, and the dimension of the vector is the total number of words to be represented. Most dimensions are 0, while only one dimension has a value of 1. Different words with a dimension of 1 are not used. If this method of representation is stored sparsely, that is, assigning a digital ID to each word based on the dimension of 1, it will be concise.

One-hot also means that no additional learning process is required under the assumption that all words are independent of each other. This maintains the orthogonality between vectors representing words and therefore has a strong discriminative ability. With maximum entropy, a support vector machine, conditional random field, and other ML algorithms, the one-hot representation has great effects on multiple aspects, such as text classification, text clustering, and part-of-speech tagging. For an application scenario of ad hoc retrieval where keyword matching plays a leading role, the bag-of-words model based on one-hot representation is still the mainstream choice.

However, one-hot representation ignores the semantic relationships between words. In addition, when representing a Vocabulary, **V**, that contains **N** words, the one-hot representation needs to construct a vector of dimension **N**. This leads to the problems of parameter explosion and data sparseness.

Another type of local representation is referred to as **bag-of-words**, or *bit vector representation*, which we introduced earlier in the chapter.

As a method of vector representation, the bag-of-words model regards the text as a collection of words, only documenting whether the words appear in the text or not but ignoring the word order and grammar in a body of text. Based on the one-hot representation of words, bag-of-words represents the text as a vector composed of 0s and 1s, and offers great support for bit operations. This method can conduct regular query processing in retrieval scenarios. Because it also maintains the orthogonality between words, it still works well for tasks such as text classification. Now, we will build a bit vector representation using a *Python ML framework* called **scikit-learn**:

```
from sklearn.feature_extraction.text import CountVectorizer

corpus = [
    'Jina is a neural search framework for neural search',
    'Jina is built with cutting edge technology called deep
    learning',
]
vectorizer = CountVectorizer(binary=True)
X = vectorizer.fit_transform(corpus)
print(X.toarray())
```

The output looks like this:

```
>>> array([[0, 0, 0, 0, 0, 1, 1, 1, 1, 0, 1, 1, 0, 0],
           [1, 1, 1, 1, 1, 0, 0, 1, 1, 1, 0, 0, 1, 1]])
```

Based on the bag-of-words (bit vector) model, the bag-of-words representation algorithm takes into account the frequency of words appearing in a body of text. Therefore, the bag-of-words encoded feature values corresponding to different words are no longer 0 or 1, but the frequency of such words appears in the body of text. Generally speaking, the more frequently a word appears in the text, the more important the word is to the text. To get the representation, you can simply put `binary=False` in the preceding implementation:

```
from sklearn.feature_extraction.text import CountVectorizer

corpus = [
    'Jina is a neural search framework for neural search',
    'Jina is built with cutting edge technology called deep
    learning',
]
vectorizer = CountVectorizer(binary=False)
X = vectorizer.fit_transform(corpus)
print(X.toarray())
```

As you can discover from the following output, the term frequency has been taken into consideration. For example, since the `neural` token occurred two times, the value of the encoded result has increased by 1:

```
>>> array([[0, 0, 0, 0, 0, 1, 1, 1, 1, 0, 2, 2, 0, 0],
           [1, 1, 1, 1, 1, 0, 0, 1, 1, 1, 0, 0, 1, 1]])
```

Last but not least, we have one of the most-used local representations, called **term frequency-inverse document frequency (tf-idf) representation**.

tf-idf is a common representation method for information retrieval and data mining. The TF-IDF value of word i in text j is as follows:

$$tf - idf_{i,j} = tf_i \times idf_i = \frac{n_{i,j}}{|d_j|} \times log \frac{|D|}{|k: t_i \in d_k|}$$

Here, $n_{i,j}$ denotes the frequency of word i appearing in the text j; $|d\_j|$ denotes the total number of words in the text; $|D|$ indicates the number of tokens in the corpus, and * $|k: t_i \in d_k|$ represents the number of documents containing the word i. By factoring in the frequency of words appearing in the text, the TF-IDF algorithm further considers the universal importance of the word in the entire body of text by calculating the IDF of the word. That is, the more frequently a word appears in the text, the less frequently it appears in other parts of the body of text. This shows that the more important the word is for the current text, the higher weight it will be given. The scikit-learn implementation of this algorithm is as follows:

```
from sklearn.feature_extraction.text import TfidfVectorizer

corpus = [
    'Jina is a neural search framework for neural search',
    'Jina is built with cutting edge technology called deep
     learning',
]
vectorizer = TfidfVectorizer()
X = vectorizer.fit_transform(corpus)
print(X.toarray())
```

The Tf-Idf weighted encoding result looks like this:

```
>>> array([[0., 0., 0., 0., 0., 0.30134034, 0.30134034,
0.21440614, 0.21440614, 0.,0.60268068, 0.60268068, 0.,
0.         ],
           [0.33310232, 0.33310232, 0.33310232, 0.33310232,
0.33310232, 0., 0. , 0.23700504, 0.23700504, 0.33310232, 0.,
0., 0.33310232, 0.33310232]])
```

Up until now, we have introduced local vector representation. In the next section, we will dive deep into a distributed vector representation, why we need it, and the commonly used algorithms.

Distributed vector representation

Although the local representation of texts has advantages in tasks such as text classification and data recall, it has the problem of data sparseness.

To be more specific, if a corpus has 100,000 distinct tokens, the dimensionality of the vector will become 100,000. Suppose we have a document that contains 200 tokens. In order to represent this document, only 200 entries of the vector out of 100,000 are non-zero. All other dimensions still get a 0 value since the tokens of the Vocabulary did not occur in the document.

This has posed great challenges to data storage and retrieval. Accordingly, a natural idea is to obtain a low-dimensional dense vector of the text, which is called a **distributed representation** of the text.

In this section, the distributed representation of single modalities, such as text, images, and audio, is first described; then, the distributed representation method of multimodal joint learning is presented. We'll also selectively introduce several important representation learning algorithms based on the modality of the data, that is, text, image, audio, and cross-modal representation learning. Let's first look at text-based algorithms.

In the following table, we have listed some selected models to encode different modalities of data:

Model	Modality	Domain	Application
BERT	Text	Dense retrieval	Text-to-text search, question answering
VGGNet	Image	Content-based image retrieval	Image-to-image search
ResNet	Image	Content-based image retrieval	Image-to-image search
Wave2Vec	Acoustic	Content-based audio retrieval	Audio-to-audio search
CLIP	Text and image	Cross-modal retrieval	Text-to-image search

Table 1.1 – Selected models that can be served as encoders for different modality of inputs

Text-based algorithms

Because text carries important information, the distributed representation of texts serves as a major function of search engines and has been extensively studied in academic works and the industry. Given the fact that we have a huge amount of unlabeled text data (such as Wikipedia), when it comes to text-based algorithms, we normally employ unsupervised pretraining on a large corpus.

Based on the belief that similar words have a similar context, Mikolov et al. proposed the *word2vec* algorithm, which includes two simple neural network models for learning: the **Continuous-Bag-of-Words** (**CBOW**) and **skip-gram** (**SG**) models.

Specifically, the CBOW model is used to derive the representation of a word, w_t, using its surrounding words, such as two words before and two words after. For example, given a sentence in a Wikipedia document, we randomly mask out one token inside this sentence. We try to predict the masked token by its surrounding tokens:

```
doc1 = 'Jina is a neural [MASK] framework'
```

In the preceding document, we masked the token search and tried to predict the vector representation of the masked token, u, by summing up the representation of surrounding tokens, v_i, and conducting the dot product between u and v_i. At training time, we'll select a token, y, to maximize the dot product:

$$argmax_y u_y^T \sum_i v_{x_i}$$

> **Important Note**
>
> It should be noted that before training, we will randomly initialize the vector values.

On the other hand, SG tries to predict the vector representations of the surrounding tokens from the current token. The difference between CBOW and SG is illustrated in *Figure 2.5*:

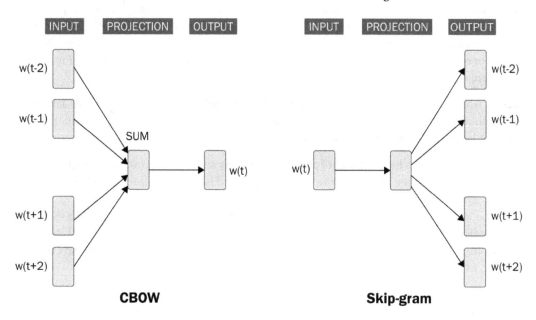

Figure 2.5 – CBOW and SG (source: Efficient estimation of word representations in vector space)

Both models are used to learn the word representation by maximizing the log-likelihood of the objective function on the entire corpus. To alleviate the burden of numerous calculations caused by the softmax function at the output layer, Mikolov et al. created two optimization methods, namely, **hierarchical softmax** and **negative sampling**. Conventional deep neural networks predict each next word as a classification task. This network must have many output classes as unique tokens. For example, when predicting the next word in the English Wikipedia, the number of classes is over 160,000. This is extremely inefficient. Hierarchical softmax and negative sampling replace the flat softmax layer with a hierarchical layer that has the words as leaves and convert the multiclass classification problem into a binary classification problem by classifying whether two tokens are a true pair (semantically similar) or a false pair (independent tokens). This greatly improves the prediction speed of word embeddings.

After pretraining, we can give a token to this word2vec model and get a so-called word embedding. This word embedding is represented by a vector. Some pretrained word2vec vectors are represented as 300D word vectors. The dimensionality is much smaller than the sparse vector space model we introduced before. So, we also refer to these vectors as dense vectors.

In algorithms such as *word2vec* and *GloVe*, the representation vector of a word generally remains unchanged after training and can be applied to downstream applications, such as named entity recognition.

However, the semantics of the same word in different contexts may vary or even have significantly different meanings. In 2019, Google announced **Bidirectional Encoder Representations from Transformers** (**BERT**), a transformer-based neural network for natural language processing. BERT uses a transformer network to represent the text and obtains the contextual information of the text through a masked language model. In addition, BERT also employs **next sentence prediction** (**NSP**) to strengthen the textual representation of relationships and has achieved good results for many textual representation tasks.

Similar to word2vec, BERT has been pretrained on the Wikipedia dataset and some other datasets, such as BookCorpus. They form a Vocabulary of above 3 billion tokens. BERT has also been trained in different languages, such as English and German, as well as on multilingual datasets.

BERT can be trained on a large amount of corpus without any annotations through the pretrain and fine-tune paradigm. During prediction, the text to be predicted is put in a well-trained network again to obtain a dynamic vector representation containing contextual information. During training, BERT replaces the words in the original text according to a certain ratio and uses the training model to make correct predictions. BERT will also add some special characters, such as `[CLS]` and `[SEP]`, to help the model correctly determine whether the two input sentences are continuous. Again, we have `doc1` and `doc2`, as follows; `doc2` is the next sentence of `doc1`:

- `doc1` = *Jina is a neural search framework*
- `doc2` = *Jina is built with cutting edge technology called deep learning*

During pretraining, we consider two documents as two sentences, and represent documents as follows:

```
doc = '[CLS] Jina is a neural [MASK] framework [SEP] Jina is
built with cutting edge technology called deep learning'.
```

After the text is input, BERT's input consists of three types of vectors, that is, **token embedding**, **segment embedding**, and **positional embedding**. The three representations are summed and then input into the subsequent transformer network. In the meantime, some tokens are being replaced by the `[MASK]` token. According to the author of the BERT paper (*BERT: Pre-training of Deep Bidirectional Transformers for Language Understanding*), around 15% of tokens are masked out (Jacob et al.).

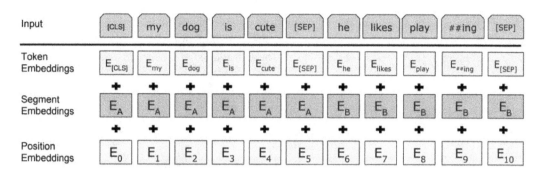

Figure. 2.6 – BERT input representation. Each input embedding is the sum of three embeddings

At the pretraining time, since we use NSP as the training objective, around 50% of the second sentences are the "true" next sentence, while another 50% of the sentences are randomly selected from the corpus, which means they're not the sentence that follows on from the first sentence. This helps us provide positive pairs and negative pairs to improve model pretraining. The objective function of BERT is to correctly predict the masked token as well as whether the next sentence is the correct one.

As was mentioned before, after pretraining BERT, we can fine-tune the model for specific tasks. The author of the BERT paper fine-tuned a pretrained model on different downstream tasks, such as question answering and language understanding, and it achieved state-of-the-art performance on 11 downstream datasets.

Vision-based algorithms

With the rapid development of the internet, information carriers on the internet are increasingly diversified and images provide a variety of visual features. Many researchers expect to encode images as vectors for representation. The most widely used model architecture for imagery analysis is called **convolutional neural network (CNN)**.

A CNN receives an image of shape (`Height`, `Width`, `Num_Channels`) as input (normally, it's a three-channel RGB image or a one-channel grayscale image). The image will be passing through one of multiple convolutional layers. This takes a kernel (or filter) and slides through the input, and the image becomes an abstracted activation map.

After one of multiple convolutional operations, the output of the activation map will be sent through a pooling layer. The pooling layer takes a small cluster of neurons in the feature map and applies max or mean operations in this cluster. This is referred to as max pooling and mean pooling. The pooling layer can significantly reduce the dimensionality of the feature map into a more compact representation.

Normally, a combination of one of multiple convolutional layers and one pooling layer is named a convolutional block. For example, three convolutional layers plus one pooling layer make a convolutional block. At the end of the convolutional block, we normally apply a flatten operation to get the vector representation of the image data.

In the following screenshot, we demonstrate a beautifully designed CNN model named VGG16. As can be seen, it consists of five convolutional blocks, each one containing two or three convolutional layers and a max pooling layer. At the end of these blocks, the activation map is flattened as a feature vector:

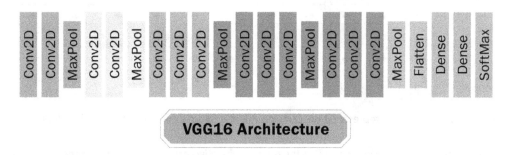

Figure. 2.7 – VGG16 consists of five convolutional blocks and produces classification results with a softmax classification head

It is worth mentioning that VGG16 is designed for ImageNet classification. So, after the activation map is flattened as a feature vector, it is connected to two fully connected layers (dense layers) and a softmax classification head.

In practice, we will remove the softmax classification head to turn this classification model into an embedding model. Given an input image, this embedding model produces a flattened feature map rather than the classified classes of the objects in the image. Besides, ResNet is a more complicated but frequently used vision feature extractor compared with VGGNet.

Apart from text and images, audio search is an important search application, for instance, to identify music from a short clip or search for music with a similar style. In the next section, we will list several deep learning models in this direction.

Acoustic-based algorithms

Given a sequence of acoustic inputs, deep learning-powered algorithms have a huge impact on the acoustic domain. For instance, they have been widely used for text-to-speech tasks. Given a piece of music as query, finding similar (or the same) music is commonly used for music applications.

One of the latest state-of-the-art algorithms trained on audio data is called **wave2vec 2.0**. Similar to BERT, wave2vec is trained in an unsupervised fashion. Taking a piece of audio data, during pretraining, wave2vec masks out parts of the audio inputs and tries to learn what has been masked out.

The major difference between wave2vec and BERT is that audio is a continuous signal with no clear segmentation into tokens. Wave2vec considers each 25 ms-long audio as a basic unit and feeds each 25 ms basic unit into a CNN model to learn a unit-level feature representation. Then, part of the input is masked out and fed into a BERT-like transformer model to predict the masked output. The training objective is to minimize the contrastive loss between the original audio and predicted audio.

It is worth mentioning that contrastive (self-supervised) pretraining is also widely used in the representation learning of text or images. For example, given an image as input, we can augment the image content a little bit to produce two views of the same image: even though these two views look different, we know they come from the same image.

This self-supervised contrastive learning has been widely used for representation learning: to learn a good feature vector given any kind of input. When applying the model to a specific domain, it is still recommended to give some labeled data to fine-tune the model with some extra labels.

Algorithms beyond text, visual, and acoustic

In real life, many kinds of information carriers exist. In addition to text, images, and speech, videos, actions, and even proteins contain a wealth of information. Therefore, many attempts have been made to obtain vector representations. Researchers at DeepMind have developed the *AlphaFold* and *AlphaFold2* algorithms. Based on traditional features, such as those of an amino acid sequence, AlphaFold algorithms can be used to obtain protein expression vectors and calculate its 3D structure in space, which greatly improves experiment efficiency in the field of protein analysis.

Moreover, in 2021, GitHub launched Copilot to help programmers with the automatic completion of code. Prior to this, OpenAI developed the *Codex* model, which was able to convert natural language into code. Based on Codex's model architecture, GitHub uses their open source TB-level code base to train the model on a large scale and completes the Copilot model to help programmers write new code. Copilot also supports the generation and completion of multiple programming languages, such as Python, JavaScript, and Go. In the search field, if we want to perform a code search or evaluate the similarity of two pieces of code, the Codex model can be employed to encode the source code into a vector representation.

The aforesaid operations mostly focus on separate encodings of text, images, or audio, so the encoded vector space may vary significantly. To map the information of different modalities to the same vector space, OpenAI researchers proposed the CLIP model, which can effectively map an image to text. Specifically, CLIP includes an image encoder and a text encoder. After inputting an image and multiple texts, CLIP encodes them at the same time and hopes to find the text most suitable for each image. By training CLIP on a large-scale dataset, CLIP can acquire an excellent representation of images and text and map them in the same vector space.

Summary

This chapter described the method of vector representation, which is a major step in the operation of search engines.

First, we introduced the importance of vector representation and how to use it, and then addressed local and distributed vector representation algorithms. In terms of distributed vector representation, the commonly used representation algorithms for text, images, and audio were covered, and common representation methods for other modalities and multimodality were summarized. Hence, we found that the dense vector representation method often entails relatively rich contextual information when compared with sparse vectors.

When building a scalable neural search system, it is important to create an encoder that can encode raw documents into high-quality embeddings. This encoding process needs to be performed fast to reduce the indexing time. At search time, it is critical to apply the same encoding process and find the top-ranked documents in a reasonable amount of time. In the next chapter, we'll utilize the ideas in this chapter and build a mental map on how to create a scalable neural search system.

Further reading

- Devlin, Jacob, et al. "Bert: Pre-training of deep bidirectional transformers for language understanding." *arXiv preprint arXiv:1810.04805* (2018).

- Simonyan, Karen and Andrew Zisserman. "Very deep convolutional networks for large-scale image recognition." *arXiv preprint arXiv:1409.1556* (2014).

- He, Kaiming et al. "Deep residual learning for image recognition." *Proceedings of the IEEE conference on computer vision and pattern recognition*. 2016.

- Schneider, Steffen, et al. "wav2vec: Unsupervised pre-training for speech recognition." *arXiv preprint arXiv:1904.05862* (2019).

- Radford, Alec et al. "Learning transferable visual models from natural language supervision." *International Conference on Machine Learning*. PMLR, 2021.

3

System Design and Engineering Challenges

Understanding **Machine Learning** (**ML**) and deep learning concepts is essential, but if you're looking to build an effective search solution powered by **Artificial Intelligence** (**AI**) and deep learning, you need production engineering capabilities as well. Effectively deploying ML models requires competencies more commonly found in technical fields such as software engineering and DevOps. These competencies are called **MLOps**. This is particularly the case for a search system that requires high useability and low latency.

In this chapter, you will learn the basics of designing a search system. You will understand core concepts such as **indexing** and **querying** and how to use them to save and retrieve information.

In this chapter, we're going to cover the following main topics in particular:

- Indexing and querying
- Evaluating a neural search system
- Engineering challenges in building a neural search system

By the end of the chapter, you will have a full understanding of the capabilities and possible difficulties to overcome when putting a neural search into production. You will be able to assess when it is useful to use neural search and which approach would be the best for your own search system.

Technical requirements

This chapter has the following technical requirements:

- A laptop with a minimum of 4 GB of RAM; 8 GB is suggested.
- Python installed, with version 3.7, 3.8, or 3.9 on a Unix-like operating system, such as macOS or Ubuntu.

The code files for the chapter are available at `https://github.com/PacktPublishing/Neural-Search-From-Prototype-to-Production-with-Jina`.

Introducing indexing and querying

In this section, you will go through two important high-level tasks to build a search system:

- **Indexing**: This is the process of collecting, parsing, and storing data to facilitate fast and accurate information retrieval. This includes adding, updating, deleting, and reading documents to be indexed.

- **Querying**: Querying is the process of parsing, matching, and ranking the user query and sending relevant information back to the user.

In a neural search system, both indexing and querying are composed of a sequence of tasks. Let's take a deep look at indexing and querying components.

Indexing

Indexing is an important process in search systems. It forms the core functionality since it assists in retrieving information efficiently. Indexing reduces the documents to the useful information contained in them. It maps the terms to the respective documents containing the information. The process of finding a relevant document in a search system is essentially identical to the process of looking at a dictionary, where the index helps you find words effectively.

Before introducing the details, we start by asking the following questions to understand where we stand:

- What are the major components of an indexing pipeline?

- What content can be indexed?

- How do we index incrementally and how do we index at speed?

If you don't know the answers to these questions, don't worry. Just keep reading!

In an indexing pipeline, we normally have three major components:

- **Preprocessor**: A preprocessor takes the raw documents into the system and performs some preprocessing tasks. For instance, if we need to index text of modality `text/plain`, we might need a tokenizer and a stemmer, as was introduced in *Chapter 1, Neural Networks for Neural Search*. If we want to index an image of modality `image/jpeg`, we might want a component to resize or transform the input image into the expected format of the neural networks. It highly depends on your task and input data.

- **Encoder**: An encoder, in a neural search system, is identical to neural networks. This neural network takes your preprocessed input as a vector representation (embeddings). After this step, each raw document composed of text, images, videos, or even DNA information should be represented as a vector of numerical values.

- **Indexer (for storage)**: An indexer, better known as **StorageIndexer**, at the indexing stage stores the vectors produced from the encoder into storage, such as memory or a database. This includes relational databases (PostgresSQL), NoSQL (MongoDB), or even better, a vector database, such as Elasticsearch.

It should be noted that every indexing task is independent. It could vary from different perspectives. For instance, if you are working on a multi-model search engine in an e-commerce context, your objective is to create a search system that can take both text and an image as a query and find the most relevant products. In this case, your indexing might have two pathways:

- Textual information should be preprocessed and encoded using text-based preprocessors and encoders.

- Likewise, image data should be preprocessed and encoded using image-based preprocessors and encoders.

You might wonder what can be indexed. Anything, as long as you have an encoder and your data can be encoded. Some common data types you can index include text, image, video, and audio. As we discussed in previous chapters, you can encode source code to build a source code search system or gene information to build a search system around that. The following figure illustrates an indexing pipeline:

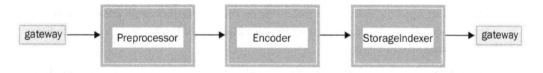

Figure 3.1 – A simple indexing pipeline takes documents as input and applies
preprocessing and encoding. In the end, save the encoded features into storage

Now, we move on to an important topic: incremental indexing. Firstly, let's discuss what incremental indexing is.

Understanding incremental indexing

Incremental indexing is a crucial feature for any search system. Given the fact that the data collection we want to index is likely to change dramatically every day, we cannot afford to index the entire data collection each time there is a small change.

In general, there are two common practices to perform an indexing task, as follows:

- **Real-time indexing**: Given any data being sent to the collection, the indexer immediately adds the document to the index.

- **Scheduled indexing**: Given any data being sent to the collection, the scheduler triggers the indexing task and performs the indexing job.

The preceding practices have their advantages and disadvantages. In real-time indexing, the user gets newly added documents immediately (if it is a match), while also consuming more system resources and potentially introducing data inconsistency. However, in the case of scheduled indexing, users don't get access to newly added results in real time but it is less error prone and easier to manage.

The indexing strategy you choose depends on your task. If the task is time sensitive, it is better to use real-time indexing. Otherwise, it is good to set up a cron job and index your data incrementally at a certain time.

Speeding up indexing

Another crucial issue when performing an indexing task in neural search is the speed of indexing. While a symbolic search system works only on textual data, the input of a neural search system could be three-dimensional (*Height * Width * ColorChannel*), such as an RGB image, or four-dimensional (*Frame * Height * Width * ColorChannel*), such as a video. This kind of data can be indexed with different modalities, which can dramatically slow down the data preprocessing and encoding process.

In general, there are several strategies that we can use to boost the indexing speed. Some of these are as follows:

- **Preprocessor**: Applying certain preprocessing on certain datasets could greatly boost your indexing speed. For instance, if you want to index high-resolution images, it's better to resize them and make them smaller.

- **GPU inference**: In a neural search system, encoding takes most of the indexing time. To be more specific, given a preprocessed document, it takes time to employ the deep neural networks to encode the document into a vector. It could be greatly improved by making use of a GPU instance for encoding. Since the GPU has much higher bandwidth memory and L1 cache, the GPU is suitable for ML tasks.

- **Horizontal scaling**: Indexing a huge amount of data on a single machine makes the process slow, but it could be much faster if we distribute data across multiple machines and perform indexing in parallel. For example, the following figure demonstrates assigning more encoders to the pipeline:

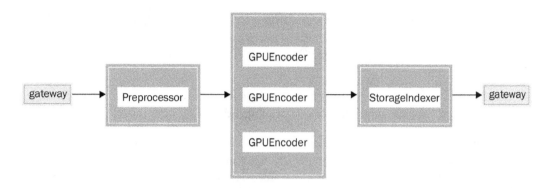

Figure 3.2 – Indexing at speed with three encoders utilizing GPU inference in parallel

It is worth mentioning that if you come from a text retrieval background, **index compression** also matters when constructing an inverted index for symbolic search. This is not exactly the same in a neural search system anymore:

- First of all, the encoder takes a document as input and encodes the content of the document into an N-dimensional vector (embeddings). Thus, we can think of the encoder itself as a compression function.

- Second, compression of the dense vectors will eventually sacrifice the quality of the vectors. Normally, larger-dimensionality vectors bring better search results since they can better represent the documents being encoded.

In practice, we need to find a balance point between dimensionality and memory usage in order to load all vectors into memory to perform a large-scale similarity search. In the next section, we will dive into the querying part, which will allow you to understand how to conduct a large-scale similarity search.

Querying

When it comes to the querying pipeline, it has a lot of components overlapping with the indexing pipeline, but with a few modifications and additional components, such as a ranker. At this stage, the input of the pipeline is a single user query. There are four major components of a typical querying task:

- **Preprocessor**: This component is similar to the preprocessor in the indexing pipeline. It takes the query document as input and applies the same preprocessors as the indexing pipeline to the input.

- **Encoder**: The encoder takes the preprocessed query document as input and produces vectors as output. It should be noted that in a cross-modal search system, your encoder from indexing might be different from the encoder at the querying step. This will be explained in *Chapter 7, Exploring Advanced Use Cases of Jina*.

- **Indexer**: This indexer, better named **SearchIndexer**, takes vectors produced from the encoder as input and conducts a large-scale similarity search over all indexed documents. This is called **Approximate Nearest Neighbor (ANN)** search. We'll elaborate more on this concept in the following section.

- **Ranker**: The ranker takes the query vector and similarity scores against each of the item within the collection, produces a ranked list in descending order, and returns results to the user.

One major difference between indexing and querying is that indexing (in most cases) is an offline task while querying is an online task. To be more concrete, when we bootstrap a neural search system and create a query, the system will return an empty list since nothing has been indexed at the moment. Before we *expose* the search system to the user, we should have pre-indexed all the documents within the data collection. This indexing is performed offline.

On the other hand, in a querying task, the user sends one query to the system and expects to get matches immediately. All the preprocessing, encoding, index searching, and ranking should be finished during the waiting time. Thus, it is an online task.

> **Important Note**
> Real-time indexing can be considered an online task.

Unlike indexing, while querying, each user sends a single document as a query to the system. The preprocessing and encoding take a very short time. On the other hand, finding similar items in the index storage becomes a critical engineering challenge that impacts the neural search system performance. Why is that?

For instance, you have pre-indexed a billion documents, and at querying time, the user sends one query to the system, the document is then preprocessed and encoded into a vector (embedding). Given the query vector, now you need to find the top N similar vectors out of 1 million vectors. How do you achieve that? Conducting similarity searches by computing distances between vectors one by one could take ages. Rather than that, we perform an ANN search.

> **Important Note**
> When we're talking about ANN search, we're considering a million-/billion-scale search. If you want to build a toy example and search across hundreds or thousands of documents, a normal linear scan is fast enough. In a production environment, please follow the selection strategy as will be introduced in the next section.

ANN search

ANN search, as defined by its name, is a trade-off between different factors: accuracy, runtime, and memory consumption. Compared with brute-force search, it ensures the running time will be accepted by the user while sacrificing precision/recall to a certain degree. How fast can it achieve? Given a billion 100-dimensional vectors, it can be fitted into a server with 32 GB memory with a 10 ms response rate. Before diving into details about ANN search, let's first take a look at the following figure:

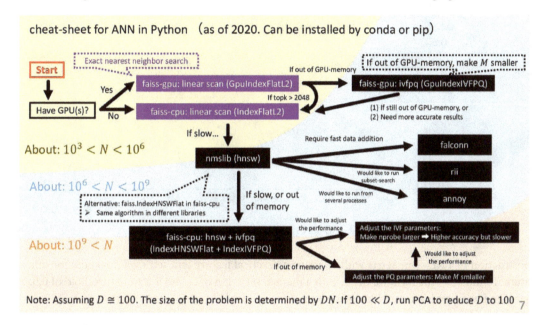

Figure 3.3 – ANN cheat sheet (source: Billion-scale Approximate Nearest Neighbor Search, Yusuke Matsui)

The preceding figure illustrates *how to select the ANN library given your search system*. In the figure, N represents the number of documents inside your *StorageIndexer*. Different numbers of N can be optimized using different ANN search libraries, such as FAISS (https://github.com/facebookresearch/faiss) or NMSLIB (https://github.com/nmslib/nmslib). Meanwhile, as you're most likely to be a Python user, the Annoy library (https://github.com/spotify/annoy) has provided a user-friendly interface with reasonable performance, and it works well enough for a million-scale vector search.

The aforementioned libraries were implemented based on different algorithms, the most popular ones being **KD-Trees**, **Locally Sensitive Hashing (LSH)**, and **Product Quantization (PQ)**.

KD-Trees follows an iterative process to construct a tree. To make the visualization easier, we suppose the data only consists of two features, *f1* (*x* axis) and *f2* (*y* axis), which looks like this:

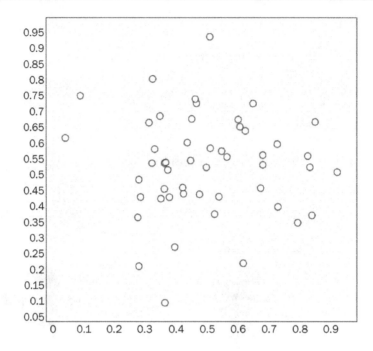

Figure 3.4 – KD-Trees, sample dataset to index

Construction of a KD-Tree starts with selecting a practical feature and setting a threshold for this feature. To illustrate the idea, we begin with a manual selection of f1 and a feature threshold of 0.5. To this end, we get a boundary like this:

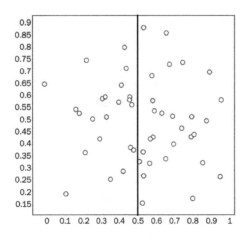 

Figure 3.5 – KD-Trees construction iteration 1

As you can see from *Figure 3.5*, the feature space has been split into two parts by our first selection of f1 with a threshold of 0.5. How is it reflected for the tree? When building the index, we're essentially creating a binary search tree. The first selection of f1 with a threshold of 0.5 became our root node. Given each data point, if the f1 is greater than 0.5, it will be placed to the right of the node. Otherwise, as shown in *Figure 3.6*, we put it to the left of the node:

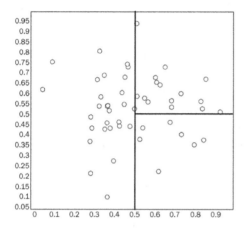
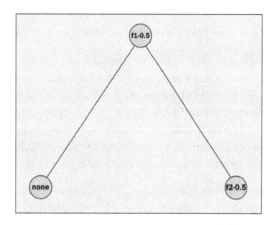

Figure 3.6 – KD-Trees construction iteration 2

We continue from the preceding tree. In the second iteration, let's define our rule as: given f1 > 0.5, select f2 with threshold 0.5. As was shown in the preceding graph, now we split the feature space again based on the new rule, and it is also reflected on our tree: we created a new node, **f2-0.5**, in the figure (the **none** node is only for visualization purpose; we haven't created this node). This is shown in the following figure:

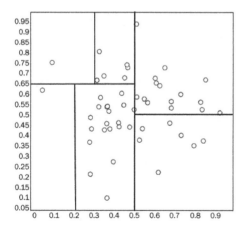
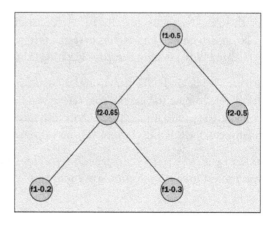

Figure 3.7 – KD-Tree construction iteration N (final iteration)

As shown in *Figure 3.7*, the entire feature space has been split into six bins. Compared with before, we added three new nodes, including two leaf nodes:

- The previous **none** was replaced by an actual node, **f2-0.65**; this node split the space of f2 based on the threshold 0.65, and it only happens when f1<0.5.

- When f2<0.65, we further split f1 by a threshold of 0.2.

- When f2>0.65, we further split f1 by a threshold of 0.3.

To this end, our tree has three leaf nodes, each leaf node can construct two bins (less/larger than the threshold), and we have six bins in total. Also, each data point can be placed into one of the bins. Then, we finish the construction of the KD-Tree. It should be noted that constructing a KD-Tree could be non-trivial since you need to consider some hyperparameters, such as how to set the threshold or how many bins we should create (or the stop criteria). In practice, there are no golden rules. Normally, the mean or median can be used to set the threshold. The number of bins could be highly dependent on the evaluation of the results and fine-tuning.

At search time, given a user query, it can be placed into one of the bins inside the feature space. We are able to compute the distance between the query and all the items within the bin as candidates for nearest neighbors. We also need to compute the minimum distance between the query and all other bins. If the distance between the query vector and other bins is greater than the distance between the query vector and the candidate for nearest neighbors, we can ignore all data points within that bin by pruning the leaf node of the tree. Otherwise, we consider the data points within that bin as candidates for nearest neighbors as well.

By constructing a KD-Tree, we do not necessarily compute the similarity between a query vector and each document anymore. Only a certain number of bins should be considered as candidates. Thus, the search time can be greatly reduced.

In practice, KD-Trees suffer from the curse of dimensionality. It is tricky to apply them to high-dimensional data because there are so many bins to search through simply because for each feature, we always create several thresholds. **Locality Sensitive Hashing (LSH)** could be a good alternative algorithm.

The basic idea behind LSH is that similar documents share the same hash code and it is designed to maximize collisions. To be more concrete: given a set of vectors, we want to have a hashing function that is able to encode similar documents into the same hashing bucket. Then, we only need to find similar vectors within the bucket (without the need to scan all the data).

Let's start with LSH index construction. At indexing time, we first need to create random planes (hyperplanes) to split the feature space into *bins*.

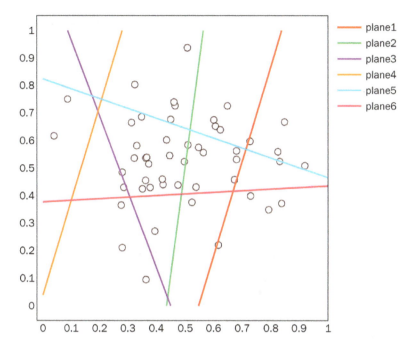

Figure 3.8 – LSH index construction with random hyperplanes

In *Figure 3.8*, we have created six hyperplanes. Each hyperplane is able to split our feature space into two bins, either left/right or up/bottom, which can be represented as binary codes (or signs): 0 or 1. This is called the index of a bin.

Let's try to get the bin index of the bottom-right bin (which has four points in the bin). The bin is located at the following points:

- Right of **plane1**, so the sign at position 0 is 1

- Right of **plane2**, so the sign at position 1 is 1

- Right of **plane3**, so the sign at position 2 is 1

- Right of **plane4**, so the sign at position 3 is 1

- Bottom of **plane5**, so the sign at position 4 is 0

- Bottom of **plane6**, so the sign at position 5 is 0

So, we can represent the bottom-right bin as 111100. If we iterate this process and annotate each bin with a bin index, we'll end up with a hash map. The keys of the hash map are the bin indexes, while the values of the hash map are the IDs of the data points within the bin.

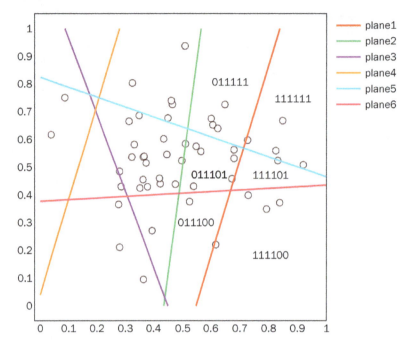

Figure 3.9 – LSH index construction with bin index

Searching on top of LSH is easy. Intuitively, given a query, you can just search all data points within its own bin, or you can search through its neighboring bins.

How do you search through its neighboring bins? Take a look at *Figure 3.9*. The bin index is represented as binary code; the neighboring bins will only have a 1-bit difference compared with its own bin index. Apparently, you can consider the difference between bin index as a hyperparameter, and search through more neighboring bins. For example, if you set the hyper parameters as 2, means you allow LSH to search through 2 neighbor bins.

To better understand this, we'll look into the Annoy implementation of LSH, namely LSH with random projection. Given a list of vectors produced by a deep neural network, we first do the following:

1. Randomly initialize a hyperplane.

2. Dot product the normal (the vector that is perpendicular to the hyperplane) against the vectors. For each vector, if the value is positive, we generate a binary code of 1, otherwise 0.

3. We generate N hyperplanes and iterate the process N times. At the end, each vector is represented by a binary vector of 0s and 1s.

4. We treat each binary code as a bucket and save all documents with the same binary code into the same bucket.

The following code block demonstrates a simple implementation of LSH with random projection:

```
pip install numpy
pip install spacy
spacy download en_core_web_md
```

We preprocess two pieces of sentences into buckets:

```
from collections import defaultdict

import numpy as np
import spacy

n_hyperplanes = 10
nlp = spacy.load('en_core_web_md')
# process 2 sentences using the model
docs = [
    nlp('What a nice day today!'),
    nlp('Hi how are you'),
]
# Get the mean vector for the entire sentence
assert docs[0].vector.shape == (300,)
# Random initialize 10 hyperplanes, dimension identical to
embedding shape
hyperplanes = np.random.uniform(-10, 10, (n_hyperplanes,
docs[0].vector.shape[0]))

def encode(doc, hyperplanes):
    code = np.dot(doc.vector, hyperplanes.T)  # dot product
vector with norm vector
    binary_code = np.where(code > 0, 1, 0)
    return binary_code

def create_buckets(docs, hyperplanes):
    buckets = defaultdict()
    for doc in docs:
        binary_code = encode(doc, hyperplanes)
        binary_code = ''.join(map(str, binary_code))
```

```
        buckets[binary_code] = doc.text
    return buckets


if __name__ == '__main__':
    buckets = create_buckets(docs, hyperplanes)
    print(buckets)
```

In this way, we map millions of documents into multiple buckets. At search time, we use the same hyperplanes to encode the search document, get the binary code, and find similar documents within the same bucket.

In the Annoy implementation, search speed is dependent on two parameters:

- search_k: This parameter is the top k elements you want to get back from the index.

- N_trees: This parameter represents the number of buckets you want to search from.

It is obvious that the search runtime highly depends on these two parameters, and the user needs to fine-tune the parameters based on their use case.

Another popular ANN search algorithm is PQ. Before we dive into PQ, it is important to understand what *quantization* is. Suppose you have a million documents to index, and you created 100 *centroids* for all documents. A **quantizer** is a function that can map a vector to a centroid. You might find the idea very familiar. Actually, the K-means algorithm is a function that can help you generate such centroids. If you do not remember, K-means works as follows:

1. Randomly initialize k centroids.

2. Assign each vector to its closest centroid. Each centroid represents a cluster.

3. Compute new centroids based on the mean of all assignments, until converge.

Once K-means converge, we get K clusters given all vectors to index. For each document to index, we create a map between the document ID and cluster index. At search time, we compute the distance query vector against the centroids and get the closest clusters, then find the closest vectors within these clusters.

This quantization algorithm has a relatively good compression ratio. You don't have to linearly scan all vectors in order to get the closest ones; you only need to scan certain clusters produced by the quantizer. On the other hand, the recall rate at searching time could be very low if the number of centroids is small. This is because there are too many edge cases that cannot be correctly distributed to the correct cluster. Also, if we simplify the set number of centroids to a large number, our K-means operations will take a long time to converge. This becomes a bottleneck at both offline indexing time and online searching time.

The basic idea behind PQ is to split high-dimensional vectors into subvectors, as illustrated in the following steps:

1. We split each vector into m subvectors.

2. For each subvector, we apply quantization. To this end, for each subvector, we have a unique cluster ID (the closest cluster of the subvector to its centroids).

3. For the full vector, we have a list of cluster IDs, which can be used as the codebook of the full vector. The dimensionality of the codebook is identical to the number of subvectors.

The following figure illustrates the PQ algorithm: given a vector, we cut it into subvectors of lower dimensionality and apply quantization. To this end, each quantized subvector gets a code:

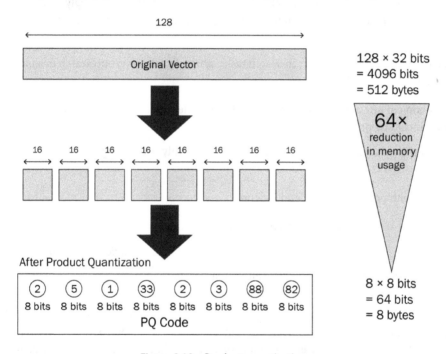

Figure 3.10 – Product quantization

At search time, again, we split the high-dimensional query vector into subvectors and generate a codebook (bucket). We compute subvector-level cosine similarity against each of the vectors inside the collection and sum up the subvector-level similarity score. We sort the final results based on the vector-level cosine similarity.

In practice, FAISS has a high-performant implementation of PQ (and beyond PQ). For more info, please refer to the documentation (https://github.com/facebookresearch/faiss/wiki).

Now we have learned two fundamental tasks, indexing and querying, for neural search. In the next section, we are going to cover neural search system evaluation to make your neural search system complete and production-ready.

Evaluating a neural search system

Evaluating the effectiveness of a neural search system is critical once you set up some baseline. By monitoring the evaluation metrics, you can immediately know how well your system performs. By diving deep into the queries, you can also conduct failure analysis and learn how to improve your system.

In this section, we will give a brief overview of the most commonly used evaluation metrics. If you want to have a more detailed mathematical understanding of this topic, we strongly recommend you go through *Evaluation in information retrieval* (`https://nlp.stanford.edu/IR-book/pdf/08eval.pdf`).

In general, given the difference between search tasks, normally we can group search evaluation into two categories:

- **Evaluation of unranked results**: These metrics are widely used in some retrieval or classification tasks, including precision, recall, and F-Score.

- **Evaluation of ranked results**: These metrics are mainly used in typical search applications given the results are ordered (ranked).

First, let's start with precision, recall, and F-Score:

- In a typical search scenario, precision is defined as follows:

  Precision = (Num of Relevant Documents Retrieved) / (Num of Retrieved Documents)

  The idea is straightforward. Suppose our search system returns 10 documents, where 7 out of 10 are relevant, then the precision would be 0.7.

  It should be noted that during evaluation, we care about the top k retrieved results. Just like the aforementioned example, we evaluate relevant documents in the top 10 results. This is normally referred to as precision, such as **Precision@10**. It also applies to other metrics we will introduce later in this section, such as Recall@10, mAP@10, and nDCG@10.

- Similarly, recall is defined as follows:

  Recall = (Num of Relevant Documents Retrieved) / (Num of Relevant Documents)

  For instance, if we search `cat` in our system, and we already know that there are 100 cat-related images being indexed, and 80 images are returned, then the recall rate is 0.8. It is an evaluation metric to measure the *completeness* of how the search system performs.

> **Important Note**
>
> Recall is the most important evaluation metric to evaluate the performance of an ANN algorithm since it depicts the fraction of true nearest neighbors found for all the queries on average.

> **Important Note**
>
> Accuracy can be a good metric for typical ML tasks, such as classification. But this is not the case for search tasks since most of the datasets in search tasks are skewed/imbalanced.

As a search system designer, you might already notice that these two numbers are trade-offs against each other: with an increased number of K, we can always expect lower precision but higher recall and vice versa. It is your decision to optimize precision or recall or optimize these two numbers as one evaluation metric, that is, F1-Score.

- F1-Score is defined as follows:

F1-Score = (2 * Precision * Recall) / (Precision + Recall)

It is a weighted harmonic mean of precision and recall. In reality, a higher recall tends to be associated with a lower precision rate. Imagine you are evaluating a ranked list and you care about the top 10 items being retrieved (and there are 10 relevant documents in the entire collection):

Document	Label	Precision	Recall
Doc1	Relevant	1/1	1/10
Doc2	Relevant	2/2	2/10
Doc3	Irrelevant	2/3	2/10
Doc4	Irrelevant	2/4	2/10
Doc5	Relevant	3/5	3/10
Doc6	Irrelevant	3/6	3/10
Doc7	Irrelevant	3/7	3/10
Doc8	Relevant	4/8	4/10
Doc9	Irrelevant	4/9	4/10
Doc10	Irrelevant	4/10	4/10

Table 3.1 – Precision recall for 10 of the top 10 documents

Table 3.1 shows the precision and recall at different levels given binary labels.

Being familiar with precision, we can now move on to calculate the **Average Precision** (**AP**). This metric will give us a better understanding of our search system's ability to sort the results of a query.

Specifically, given the preceding-ranked list, aP@10 is as follows:

```
aP@10 = (1/1 + 2/2 + 3/5 + 4/8) / 10 = 0.31
```

Note that only the precision of relevant documents is taken into consideration when computing aP.

Now, the aP has been calculated against one specific user query. However, to give a more robust search system evaluation, we want to evaluate the performance of a collection of user queries as a test set. This is called **Mean Average Precision (mAP)**. For each query, we calculate aP@k, then we average all the aPs over a set of queries to get the mAP score.

mAP is one of the most important search system evaluation metrics given an ordered rank list. To conduct mAP evaluation on your search system, normally you have to follow these steps:

1.   Compose a list of queries to have a good representation of users' information needs. The number is dependent on your situation, such as 50, 100, or 200.

2.   If your documents already have labels that indicate the degree of relevance, use the labels directly to compute aP per query. If your documents do not contain any relevant information against each query, we need expert annotation or pooling to access relevant degrees.

3.   Compute mAP over a list of queries by taking the average of the aP. As was mentioned previously, if you do not have a relevant assessment for ranked documents, one common technique is called **pooling**. It requires us to set up multiple search systems (such as three) for testing. Given each query, we collect the top K documents returned by each of these three search systems. A human annotator judges the degree of relevance of all 3 * K documents. Afterward, we consider all documents outside this pool as irrelevant, while all documents inside this pool are relevant. Then, the search results can be evaluated on top of the pools.

At this point, even though mAP is evaluating a ranked list, the nature of the definition of precision still neglects some of the nature of a search task: precision is evaluated based on binary labels, either relevant or irrelevant. It does not reflect the *relatedness* of the query against documents. **Normalized Discounted Cumulative Gain (nDCG)** is used for evaluating search system performance over the degree of relatedness.

nDCG can have multiple levels of rating for each document, such as *irrelevant*, *relevant*, or *highly relevant*. In this case, mAP does not work anymore.

For instance, given three degrees of relevance (irrelevant, relevant, and highly relevant), these differences in degrees of relevance can be represented as information gain that users can obtain by getting each document. For highly relevant documents, the gain could be assigned a value of *3*, relevant could be *1*, and not relevant could be set to *0*. Then, if a highly relevant document is ranked higher than documents that are not relevant, the user could cumulate more *gain*, which is referred to as **Cumulative Gain (CG)**. The following table shows the information gain we get per document given the top 10 ranked documents produced by a search system:

Document	Label	Gain
Doc1	Highly Relevant	3
Doc2	Relevant	1
Doc3	Irrelevant	0
Doc4	Irrelevant	0
Doc5	Relevant	1
Doc6	Irrelevant	0
Doc7	Irrelevant	0
Doc8	Highly Relevant	3
Doc9	Irrelevant	0
Doc10	Irrelevant	0

Table 3.2 – Top 10 documents with information gain

In the preceding table, the system returned the top 10 ranked documents to the user. Based on the relevance degree, we assign 3 as the gain to highly relevant documents, 1 as the gain to relevant documents, and 0 as the gain to irrelevant documents. The CG is the sum of all gains in the top 10 documents, such as the following:

```
CG@10 = 3 + 1 + 1 + 3 = 8
```

But think about the nature of a search engine: the users scan the ranked list from top to bottom. So, by nature, the top-ranked documents should have more gains than the documents ranked lower so that our search system will try to rank highly relevant documents in higher positions. So, in practice, we will penalize the gain by the position. See the following example:

Document	Label	Gain	Discounted Gain
Doc1	Highly Relevant	3	3
Doc2	Relevant	1	1/log2
Doc3	Irrelevant	0	0
Doc4	Irrelevant	0	0
Doc5	Relevant	1	1/log5
Doc6	Irrelevant	0	0
Doc7	Irrelevant	0	0
Doc8	Highly Relevant	3	3/log8
Doc9	Irrelevant	0	0
Doc10	Irrelevant	0	0

Table 3.3 – Top 10 documents of gain and discounted gain

In the preceding table, given the gain of a document and its ranked position, we penalize the gain a little bit by dividing the gain by a factor. In this case, it's the logarithm of the ranked position. The sum of the gain is called the **Discounted Cumulative Gain (DCG)**:

```
DCG@10 = 3 + 1/log2 + 1/log5 + 3/log8 = 6.51
```

Before we start to compute nDCG, it's important to understand the concept of ideal DCG. It simply means the best-ranking result we could achieve. In the preceding case, if we look at the top 10 positions, ideally the ranked list should contain all highly relevant documents with a gain of 3. So, the iDCG should be as follows:

```
iDCG@10 = 3 + 3/log2 + 3/log3 + 3/log4 + 3/log5 + 3/log6 + 3/log7 +
3/log8 + 3/log9 + 3/log10 = 21.41
```

In the end, the final nDCG is as follows:

```
nDCG = DCG/iDCG
```

In our preceding example, we have the following:

```
nDCG@10 = 6.51/21.41 = 0.304
```

It is worth mentioning that even though nDCG is well suited to evaluating a search system that reflects the degree of relevance, the *relatedness* itself is biased toward different factors, such as search context and user preference. It is non-trivial to perform such an evaluation in a real-world scenario. In the next chapter, we will dive into the details of such challenges and briefly introduce how to resolve them.

Engineering challenges of building a neural search system

Now, you would have noticed that the most important building blocks of the neural search system are the encoder and indexer. The quality of encoding posts has a direct impact on the final search result, while the speed of the indexer determines the scalability of your neural search system.

Meanwhile, this is still not enough to make your neural search system ready to use. Many other topics need to be taken into consideration as well. The first question is: does your encoder (neural model) have the same distribution as your data? For new people coming into the neural search system world who are using a pretrained deep neural network, such as ResNet trained on ImageNet, it is trivial to quickly set up a search system. However, if your target is to build a neural search system on a specific domain, let's say a fashion product image search, it is not going to produce satisfying results.

One important topic before we really start creating an encoder and setting up our search system is applying transfer learning to your dataset and evaluating the match results. This means taking a pretrained deep learning model, such as ResNet, chopping off the head layer, freezing the weights of the pretrained model, and attaching a new embedding layer to the end of the model, then training it on your new dataset on your domain. This could greatly boost search performance.

Apart from that, in some vision-based search systems, purely relying on the encoder might not be sufficient. For instance, a lot of vision-based search systems rely heavily on an object detector. Before sending the full image into the encoder, it should be sent to the object detector first and the meaningful part of the image extracted (and remove the background noise). This is likely to improve the embedding quality. Meanwhile, some vision-based classification models could also be employed to enrich the search context as a hard filter. For instance, if you are building a neural search system that allows people to search similar automobiles given an image as a query, a pretrained brand classifier could be useful. To be more concrete, you pretrain an automobile brand classifier to *recognize* different car brands based on images and apply the recognition to the indexing and searching pipeline. Once a vision-based search is complete, you can apply the recognized brand as a hard filter to filter out cars of other brands.

Similarly, for text-based search, when a user provides keywords as a query, directly applying an embedding-based similarity search might be insufficient. For example, you can create a **Named Entity Recognition** (**NER**) module in your indexing and querying pipeline to enrich the metadata.

For web-based search engines such as Google, Bing, or Baidu, it is very common to see query automatic completion. It might be also very interesting to add a deep neural network-powered keyword extraction component to your indexing and searching pipeline to use a similar user experience.

To summarize, to build a production-ready neural search system, it is very challenging to design a feature-complete indexing and querying pipeline, given the fact that search is such a complex task. Designing such a system is already challenging, let alone engineering the infrastructure. Luckily, Jina can already help you with most of the most challenging tasks.

Summary

In this chapter, we have discussed the fundamental tasks to build a neural search system, which are the indexing and querying pipelines. We looked into both of them and introduced the most challenging part, such as encoding and indexing.

You should have basic knowledge of the basic building blocks of indexing and querying, such as preprocessing, encoding, and indexing. You should also notice that the quality of the search results highly depends on the encoder, while the scalability of the neural search system highly depends on the indexer and the most popular algorithms behind the indexer.

As you need to build a production-ready search system, you will realize that purely relying on the basic building blocks is not enough. As a search system is complex to implement, it is always needed to design and add your own building blocks to the indexing and querying pipeline, in order to bring better search results.

In the next chapter, we will start to introduce Jina, the most popular framework that helps you engineer a neural search system. You will realize that Jina has tackled the most difficult problems for you and could make your life as a neural search system engineer/scientist much easier.

Part 2: Introduction to Jina Fundamentals

In this part, you will learn about what Jina is and its basic components. You will understand its architecture and how can it be used to develop deep-learning searches on the cloud. The following chapters are included in this part:

- *Chapter 4, Learning Jina's Basics*
- *Chapter 5, Multiple Search Modalities*

4

Learning Jina's Basics

In the previous chapter, we learned about neural search, and now we can start thinking about how to work with it and the steps we'll need to take to implement our own search engine. However, as we saw in previous chapters, in order to implement an end-to-end search solution, time and effort will be needed to gather all of the resources required. This is where Jina can help as it will take care of many of the necessary tasks, letting you focus on the design of your implementation.

In this chapter, you will understand the core concepts of Jina: **Documents**, **DocumentArrays**, **Executors**, and **Flow**. You will see each of them in detail and understand their overall design and how they connect.

We're going to cover the following main topics:

- Exploring Jina
- Documents
- DocumentArrays
- Executors
- Flow

By the end of this chapter, you will have a solid understanding of idioms in Jina, what they are, and how to use them. You will use this knowledge later to build your own search engine for any type of modality.

Technical requirements

This chapter has the following technical requirements:

- A laptop with a minimum of 4 GB of RAM, ideally 8 GB
- Python 3.7, 3.8, or 3.9 installed on a Unix-like operating system, such as macOS or Ubuntu

Exploring Jina

Jina is a framework that helps you build deep learning search systems on the cloud using state-of-the-art models. Jina is an infrastructure that allows you to focus only on the areas that you are interested in. In this way, you don't need to be involved in every aspect of building a search engine. This involves everything from pre-processing your data to spinning up microservices if needed. Another good thing about neural search is that you can search for any kind of data regardless of type. Here are some examples of how you can search using different data types:

- Image-to-image search
- Text-to-image search
- Question answering search
- Audio search

Building your own search engine can be very time-consuming, so one of the core goals of Jina is reducing the time you would need if you were going to build one from scratch. Jina is designed in a layered way that lets you focus only on the specific parts that you need, letting the rest of the infrastructure be handled in the background. So, for example, you could use pre-trained **Machine Learning** (**ML**) models directly instead of building them yourself.

Since we live in the era of cloud computing, it makes sense to leverage the power that decentralized work can offer, so it is useful to design your solution to be distributed on the cloud, and features such as **sharding**, **asynchronizing**, and **REST** are fully integrated and work out of the box.

As we have already said, another way that Jina helps you reduce the time and effort needed while building a search engine is by using the latest state-of-the-art ML models. You take advantage of this in one of two ways:

- Using one of Jina's plug-and-play models
- Developing your own model from scratch for when you have a specialized use case or if there is still no model available on Jina Hub

With these options, you can choose between having a pre-defined model or implementing your own if your needs are not covered.

As you can imagine, all of this means that there are a lot of components working in the background. The more you learn, the more power you will have over your application, but to start, you will need to understand the basic components of Jina, which we will discuss in the following sections:

- Documents
- DocumentArrays

- Executors
- Flows

Documents

In Jina, **Documents** are the most basic data type you can work with. They are the data you want to use and can be used for indexing and/or querying. They can be made with whatever data type you require, such as text, gifs, PDF files, 3D meshes, and so on.

We will use Documents to index and query, but since Documents can be of any type and size, it's likely that we will need to divide them before use.

As an analogy, think of a Document as a chocolate bar. There are several types of chocolate: white, dark, milk, and so on. Likewise, a Document can be of several types, such as audio, text, video, a 3D mesh, and so on. Also, if we have a big chocolate bar, we will probably divide it into smaller pieces before eating it. Accordingly, if we have a big Document, we should divide it into smaller pieces before indexing.

This is how a Document looks in Python code:

```
from jina import Document
document = Document()
```

As you can see, all you need to create a Document is to import it from Jina and create it as you would any other object in Python. This is a very basic example, but in real life, you will have more complex cases, so we will need to add some attributes, which we will see next.

Document attributes

Each Document can have different attributes that belong to four main categories:

- **Content**: This refers to the actual content of your Document. For example, the text or its embedded vector.
- **Meta**: This is information about the Document itself. For example, its ID and whether it has any tags.
- **Recursive**: This tells us how the Document is divided. For example, its matches or if it was divided into any chunks.
- **Relevance**: This refers to the relevance of the Document, such as its score.

These categories consist of various attributes, which are listed in the following table:

Category	Attributes
Content attributes	`.buffer, .blob, .text, .uri, .content, .embedding`
Meta attributes	`.id, .parent_id, .weight, .mime_type, .content_type,` `.tags, .modality`
Recursive attributes	`.chunks, .matches, .granularity, .adjacency`
Relevance attributes	`.score, .evaluations`

Table 4.1 – Document categories and their attributes

We will see later in more detail what each of those attributes are, but first, let's see how to set them.

Setting and unsetting attributes

The attributes in *Table 4.1* are the possible attributes we can use with our Document. Let's say we want our Document to have the text `hello world`. We can do this by setting its `text` attribute like so:

```
from jina import Document
document = Document()
document.text = 'hello world'
```

And if we want to unset it, we can do so as follows:

```
document.pop('text')
```

In plenty of real-world cases, we will need to work with multiple attributes, and it is also possible to unset several of these at once:

```
document.pop('text', 'id', 'mime_type')
```

Accessing nested attributes from tags

In Jina, each Document contains tags that hold a map-like structure that can map arbitrary values:

```
from jina import Document
document = Document(tags={'dimensions': {'height': 5.0,
'weight': 10.0}})
document.tags['dimensions'] # {'weight': 10.0, 'height': 5.0}
```

If you want to access the nested fields, you can do so by using the attribute name with the symbol __
interlaced. For example, if you would like to access the weight tag, you should do the following:

```
from jina import Document
document = Document(tags={'dimensions': {'height': 5.0,
'weight': 10.0}})
Document.tags__dimensions__weight #10
```

Constructing a Document

To construct a Document, you need to fill it with attributes, so let's take a look at them.

Content attributes

Each Document needs to contain some information about itself, ranging from raw binary content to
text info. We can see the details that a Document can have in the following table:

Attribute	Description
doc.buffer	The raw binary content of the Document
doc.blob	The ndarray of the image/audio/video Document
doc.text	The text info of the Document
doc.uri	A Document URI could be a local file path, a remote URL that starts with http or https, or a data URI scheme
doc.content	This can be any of the previous attributes (buffer, blob, text, uri)
doc.embedding	The embedding ndarray of the Document

Table 4.2 - Content attributes

There are two ways you can assign a *content* type to your Document. If you know exactly what type it
is, you can assign it explicitly with the text, blob, buffer, or uri attributes. If you don't know
the type, you can use .content, which will automatically assign a type to your Document based
on what it's most likely to be. See this, for example:

```
from jina import Document
import numpy as np

document1 = Document(content='hello world')
document2 = Document(content=b'\f1')
document3 = Document(content=np.array([1, 2, 3]))
```

```
document4 = Document(content=
'https://static.jina.ai/logo/core/notext/light/logo.png')
```

In this example, the following applies:

- document1 will have a field of text.

- document2 will have a field of buffer.

- document3 will have a field of blob.

- document4 will have a field of uri.

The content will be automatically assigned to any one of the text, buffer, blob, or uri fields. The id and mime_type attributes are auto-generated when not set explicitly. This means that you can specify explicitly the ID and type (mime_type) of your document, otherwise it will be autogenerated.

Exclusivity of doc.content

In Jina, each Document can only contain one type of content: text, buffer, blob, or uri. Setting text first and then setting uri will clear the text field.

In the following figure, you can see the different types that content can have, as well as the fact that each Document can only have one type.

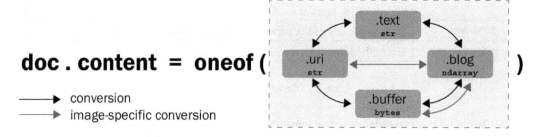

Figure 4.1 – Possible content types in a Document

Let's see how we would set the content attribute of a Document in code:

```
doc = Document(text='hello world')
doc.uri = 'https://jina.ai/' #text field is cleared, doc
#has uri field now
assert not doc.text  # True


doc = Document(content='https://jina.ai')
assert doc.uri == 'https://jina.ai'  # True
```

```
assert not doc.text  # True
doc.text = 'hello world' #uri field is cleared, doc has
#text field now

assert doc.content == 'hello world'  # True
assert not doc.uri  # True
```

You can see how you can set every type of attribute in a Document, but if you assign different values to a single Document, only the last attribute will be valid.

Conversion of doc.content

Now that you've seen the different possible attributes in Jina, you might be thinking that sometimes it'd be useful to convert one type of doc.content to another. For example, if you had a Document and its path (uri), but you needed it in a text format, you could use one of these pre-made conversion functions to easily switch the content type:

```
doc.convert_buffer_to_blob()
doc.convert_blob_to_buffer()
doc.convert_uri_to_buffer()
doc.convert_buffer_to_uri()
doc.convert_text_to_uri()
doc.convert_uri_to_text()
doc.convert_image_buffer_to_blob()
doc.convert_image_blob_to_uri()
doc.convert_image_uri_to_blob()
doc.convert_image_datauri_to_blob()
```

As you can see, all of these methods will help you transform your data from one type to another, but all of these types will need to be transformed into vector embeddings . Let's see what exactly an embedding is and why we use them in neural search.

Setting an embedding attribute

An embedding is a high-dimensional representation of a Document, and it's a key element in neural search. Embeddings are representations of your data in a vector format. This is why neural search can be used for any kind of data regardless of its type (image, audio, text, and so on). The data will be transformed into vectors (embeddings), and those vectors are the ones that will be used in neural search. Therefore, the type doesn't matter as neural search ends up working only with vectors.

Since we are working with vectors, it is useful to work with already-established libraries that have extensive support for embeddings, such as NumPy, so you can, for example, assign any NumPy ndarray as a Document's embedding and then use the flexibility that those libraries provide:

```
import numpy as np
from jina import Document
d1 = Document(embedding=np.array([1, 2, 3]))
d2 = Document(embedding=np.array([[1, 2, 3], [4, 5, 6]]))
```

Meta attributes

Apart from content attributes, you can also have meta attributes:

Attribute	Description
doc.tags	Used to store the meta-information of the Document
doc.id	A hexdigest that represents a unique Document ID
doc.parent_id	A hexdigest that represents the Document's parent ID
doc.weight	The weight of the Document
doc.mime_type	The mime type of the Document
doc.content_type	The content type of the Document
doc.modality	An identifier of the modality of the Document such as an image, text, and so on

Table 4.3 – Meta attributes

To create your Document, you can assign multiple attributes in its constructor as shown here:

```
from jina import Document
document = Document(uri='https://jina.ai',
                mime_type='text/plain',
                granularity=1,
                adjacency=3,
                tags={'foo': 'bar'})
```

Constructing a Document from dictionary or a JSON string

There's also the option to construct your Document directly from a Python dictionary or JSON string. If you have the information of your Document already stored in those formats, you can conveniently create the Document using the following example:

```
from jina import Document
import json
```

```
doc = {'id': 'hello123', 'content': 'world'}
doc1 = Document(d)


doc = json.dumps({'id': 'hello123', 'content': 'world'})
doc2 = Document(d)
```

Parsing unrecognized fields

If the fields in dictionary/a JSON string are not recognized, they are automatically put into the `document.tags` field. As shown in the following example, `foo` is not a defined attribute (*Table 4.3*), so it will be automatically parsed into the `tags` field:

```
from jina import Document
doc1 = Document({'id': 'hello123', 'foo': 'bar'})
```

You can use `field_resolver` to map external field names to Document attributes:

```
from jina import Document
doc1 = Document({'id': 'hello123', 'foo': 'bar'},
field_resolver={'foo': 'content'})
```

Constructing a Document from other Documents

In case you want to duplicate a Document, the following are ways to do so:

- **Shallow copy**: Assigning a Document object to another Document object will make a shallow copy:

  ```
  from jina import Document


  doc = Document(content='hello, world!')
  doc1 = doc
  assert id(doc) == id(doc1)  # True
  ```

- **Deep copy**: To make a deep copy, use `copy=True`:

  ```
  doc1 = Document(doc, copy=True)
  assert id(doc) == id(doc1)  # False
  ```

- **Partial copy**: You can partially update a Document according to another source Document:

```
from jina import Document
s = Document(
    id='🐍',
    content='hello-world',
    tags={'a': 'b'},
    chunks=[Document(id='🐢')],
)
d = Document(
    id='🦎',
    content='goodbye-world',
    tags={'c': 'd'},
    chunks=[Document(id='🐠')],
)
# only update `id` field
d.update(s, fields=['id'])
# update all fields. `tags` field as `dict` will be merged.
d.update(s)
```

You can use any of the three preceding methods to copy a Document.

Constructing a Document from file types such as JSON, CSV, ndarray, and others

The jina.types.document.generators module lets you construct Documents from common file types such as JSON, CSV, ndarray, and text files.

The following functions will create a generator of Documents, where each Document object corresponds to a line/row in the original format:

Import Method	Description
from_ndjson()	This function yields a Document from a line-based JSON file. Each line is a Document object.
from_csv()	This function yields a Document from a .csv file. Each line is a Document object.
from_files()	This function yields a Document from a glob file. Each file is a Document object.
from_ndarray()	This function yields a Document from an ndarray. Each row (depending on the axis) is a Document object.
from_lines()	This function yields a Document from lines, of JSON, and CSV.

Table 4.4 – Python methods for constructing Documents

Using a generator is sometimes less memory-intensive, as it does not load/build all Document objects in one go.

Now you have learned what a Document is and how to create one. You can create it either by filling it with individual bits of content, or by copying from a JSON file if you already have one.

DocumentArray

Another powerful concept in Jina is the **DocumentArray**, which is a list of Document objects. If you need multiple Documents, you can group them all together in a list using DocumentArray. You can use a DocumentArray as a regular list in Python with all of the usual methods, such as `insert`, `delete`, `construct`, `traverse`, and `sort`. The DocumentArray is a first-class citizen to an Executor, serving as its input and output. We will talk about Executors in the next section, but for now, think of them as the way Jina processes Documents.

Constructing a DocumentArray

You can construct, delete, insert, sort, and traverse a `DocumentArray` like a Python list. You can create these in different ways:

```
from jina import DocumentArray, Document
documentarray = DocumentArray([Document(), Document()])
```

```
from jina import DocumentArray, Document
documentarray = DocumentArray((Document() for _ in range(10))
```

```
from jina import DocumentArray, Document
documentarray1 = DocumentArray((Document() for _ in range(10)))
documentarray2 = DocumentArray(da)
```

Just like a normal Document, the DocumentArray also supports different methods as follows:

Category	Attributes
Python list-like interface	`__getitem__`, `__setitem__`, `__delitem__`, `__len__`, insert, append, reverse, extend, `__iadd__`, `__add__`, `__iter__`, clear, sort, shuffle, sample
Persistence	save, load

Neural search operations	match, visualize
Advanced getters	get_attributes, get_attributes_with_docs, traverse_flat, traverse

Table 4.5 – DocumentArray attributes

Persistence via save()/load()

Of course, there will be cases where you want to save the elements of your DocumentArray for further processing, and you can save all elements in a DocumentArray in two ways:

- In JSON line format
- In binary format

To save it in JSON line format, you can do the following:

```
from jina import DocumentArray, Document

documentarray = DocumentArray([Document(), Document()])
documentarray.save('data.json')
documentarray1 = DocumentArray.load('data.json')
```

And to store it in binary format, which is much faster and yields smaller files, you can do the following:

```
from jina import DocumentArray, Document

documentarray = DocumentArray([Document(), Document()])
documentarray.save('data.bin', file_format='binary')
documentarray1 = DocumentArray.load('data.bin', file_format='binary')
```

Basic operations

Like with any other object, you can perform basic operations on a DocumentArray. This includes the following:

- Accessing elements
- Sorting elements
- Filtering elements

Let's learn about these in detail.

Accessing elements

You can access a Document in the DocumentArray via an index, ID, or slice indices, as shown here:

```
from jina import DocumentArray, Document

documentarray = DocumentArray([Document(id='hello'),
Document(id='world'), Document(id='goodbye')])
documentarray[0]
# <jina.types.document.Document id=hello at 5699749904>

documentarray['world']
# <jina.types.document.Document id=world at 5736614992>

documentarray[1:2]
# <jina.types.arrays.document.DocumentArray length=1 at
# 5705863632>
```

Feel free to use to any variation of these options depending on your use case scenario.

Sorting elements

Because DocumentArray is a subclass of MutableSequence, you can use the built-in Python function sort to sort elements in a DocumentArray. For example, if you want to sort elements in-place (without making copies), and use the tags[id] value in a descending manner, you can do the following:

```
from jina import DocumentArray, Document

documentarray = DocumentArray(
    [
        Document(tags={'id': 1}),
        Document(tags={'id': 2}),
        Document(tags={'id': 3})
    ]
)

documentarray.sort(key=lambda d: d.tags['id'],
reverse=True)
print(documentarray)
```

The preceding code would print the following:

```
<jina.types.arrays.document.DocumentArray length=3 at
5701440528>

{'id': '6a79982a-b6b0-11eb-8a66-1e008a366d49', 'tags': {'id':
3.0}},
{'id': '6a799744-b6b0-11eb-8a66-1e008a366d49', 'tags': {'id':
2.0}},
{'id': '6a799190-b6b0-11eb-8a66-1e008a366d49', 'tags': {'id':
1.0}}
```

Filtering elements

You can use Python's built-in `filter` function to filter elements in a `DocumentArray` object:

```
from jina import DocumentArray, Document

documentarray = DocumentArray([Document() for _ in range(6)])
for j in range(6):
    documentarray[j].scores['metric'] = j
for d in filter(lambda d: d.scores['metric'].value > 2,
documentarray):
    print(d)
```

This would print the following:

```
{'id': 'b5fa4871-cdf1-11eb-be5d-e86a64801cb1', 'scores':
{'values': {'metric': {'value': 3.0}}}}
{'id': 'b5fa4872-cdf1-11eb-be5d-e86a64801cb1', 'scores':
{'values': {'metric': {'value': 4.0}}}}
{'id': 'b5fa4873-cdf1-11eb-be5d-e86a64801cb1', 'scores':
{'values': {'metric': {'value': 5.0}}}}
```

You can also build a `DocumentArray` object from the filtered results as follows:

```
from jina import DocumentArray, Document

documentarray = DocumentArray([Document(weight=j) for j in
range(6)])
```

```
documentarray2 = DocumentArray(d for d in documentarray if
d.weight > 2)
print(documentarray2)
```

This would print the following result:

```
DocumentArray has 3 items:
{'id': '3bd0d298-b6da-11eb-b431-1e008a366d49', 'weight': 3.0},
{'id': '3bd0d324-b6da-11eb-b431-1e008a366d49', 'weight': 4.0},
{'id': '3bd0d392-b6da-11eb-b431-1e008a366d49', 'weight': 5.0}
```

At this point, you have learned how to create Documents and DocumentArrays that store multiple Documents as a list. But what can you actually do with these? How can you process them for use in neural search? This is where Executors come into the picture. Let's talk about them in the following section.

Executors

The **Executor** represents the processing component in a Jina Flow. It performs a single task on a Document or DocumentArray. You can think of an Executor as the logical part of Jina. Executors are the ones that will perform tasks of all kinds on a Document. For example, you could have an Executor for extracting text from a PDF file, or for encoding audio for your Document. They handle all of the algorithmic tasks in Jina.

Since Executors are one of the main parts of Jina, and they are the ones that perform all the algorithmic tasks, it would be very useful for you to make them in a way that means they could be easily shared with other people, so that others can re-use your work. Similarly, you could use prebuilt Executors made by someone else in your own code. This is in fact possible because Executors are easily available in a marketplace, which in Jina is called Jina Hub (https://hub.jina.ai/). There you can browse between various Executors that solve different tasks, and you can just select the one that is useful to you and use it in your code. Of course, it's possible that the Executor for the task you need to do has not already been built in Jina Hub, in which case you'll need to create your own Executor. You can do this easily in Jina Hub. Let's take a deep dive into how to do that.

Creating an Executor

To create an Executor, it's best to use Jina Hub, which will generate a wizard to guide you through the process. To start this process, open a console and write the following command:

```
jina hub new
```

This will trigger a wizard that will guide you through the creation of the Executor and ask you for some details about it:

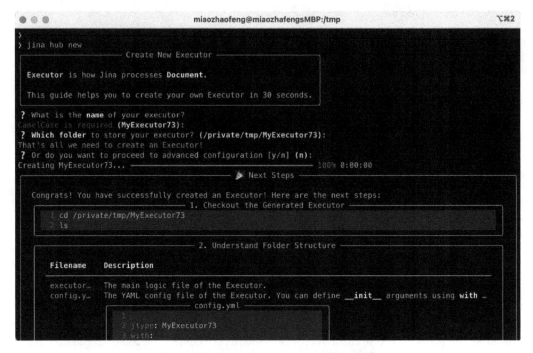

Figure 4.3 – Creating an Executor via the CLI

After going through the wizard, your Executor will be ready. Now, let's learn about Executors in more detail.

Executors process DocumentArrays in-place via functions decorated with `@requests`. We add this decorator to whatever functions we want in our Executors. When creating an Executor, three principles should be kept in mind:

- It should be a subclass of the `jina.Executor` class.
- It must be a bag-of-functions with the state `shared`. It can contain an arbitrary number of functions with arbitrary names.
- Functions decorated by `@requests` will be invoked according to their `on=` endpoint. We will see different cases of what those endpoints could be in the following example.

Here is a very basic Executor in Python to help you understand this last concept better:

```
from jina import Executor, requests

class MyExecutor(Executor):
```

```
@requests
def foo(self, **kwargs):
    print(kwargs)
```

The name of your Executor can be whatever you wish, but the important thing to remember is that every new Executor should be a subclass of jina.Executor.

Constructor

You don't need to implement the constructor (__init__) if your Executor does not contain initial states, but if your Executor has __init__, it needs to carry **kwargs in the signature and call super().__init__(**kwargs) into the body:

```
from jina import Executor

class MyExecutor(Executor):
    def __init__(self, foo: str, bar: int, **kwargs):
        super().__init__(**kwargs)
        self.bar = bar
        self.foo = foo
```

Method decorator

The @requests decorator defines when a function will be invoked. It has the on= keyword, which defines the endpoint. We haven't talked about Flow yet. We will do so in the next section, but for now, think of Flow as a manager. The @requests decorator sends information to Flow whenever our Executor needs to be called. This is to communicate to Flow when the function will be called and at which endpoint.

You can use the decorator like this:

```
from jina import Executor, Flow, Document, requests
class MyExecutor(Executor):
    @requests(on='/index')
    def foo(self, **kwargs):
        print(f'foo is called: {kwargs}')

    @requests(on='/random_work')
    def bar(self, **kwargs):
        print(f'bar is called: {kwargs}')
```

```
f = Flow().add(uses=MyExecutor)
with f:
    f.post(on='/index', inputs=Document(text='index'))
    f.post(on='/random_work',
    inputs=Document(text='random_work'))
    f.post(on='/blah', inputs=Document(text='blah'))
```

In this example, we have three endpoints:

- `on='/index'`: This endpoint will trigger the `MyExecutor.foo` method.

- `on='/random_work'`: This endpoint will trigger the `MyExecutor.bar` method.

- `on='/blah'`: This endpoint will not trigger any methods, as no function is bound to `MyExecutor.blah`.

Executor binding

Now that we have seen how to create Executors and learned about the `@requests` decorator, you might be wondering what types of binding you can use with `@requests`.

Default binding

A class method decorated with plain `@requests` is the default handler for all endpoints. This means that it is the fallback handler for endpoints that are not found. Let's see one example:

```
from jina import Executor, requests


class MyExecutor(Executor):
    @requests
    def foo(self, **kwargs):
        print(kwargs)

    @requests(on='/index')
    def bar(self, **kwargs):
        print(kwargs)
```

In this example, two functions were defined:

- `foo`
- `bar`

Here the `foo` function becomes the default method since it has no `on=` keyword. If we were now to use an unknown endpoint, such as `f.post(on='/blah', ...)`, it would invoke `MyExecutor.foo` since there is no `on='/blah'` endpoint.

Multiple bindings

To bind a method with multiple endpoints, you can use `@requests(on=['/foo', '/bar'])`. This allows either `f.post(on='/foo', ...)` or `f.post(on='/bar', ...)` to invoke the function.

No binding

A class with no `@requests` binding plays no part in the a Flow. The request will simply pass through without any processing.

Now you know what an Executor is and why it is useful to share them with other developers. You have also learned where to find already-published Executors and how to publish your own. Let's see now how to put together the concepts you've learned so far.

Flow

Now that you know what Documents and Executors are and how to work with them, we can start to talk about **Flow**, one of the most important concepts in Jina.

Think of Flow as a manager in Jina; it takes care of all the tasks that will run on your application and will use Documents as its input and output.

Creating a Flow

The creation of a Flow in Jina is very easy and works just like any other object in Python. For example, this is how you would create an empty Flow:

```
from jina import Flow
f = Flow()
```

In order to use a Flow, it's best to always open it as a context manager, just like you would open a file in Python, by using the `with` function:

```
from jina import Flow
f = Flow()
with f:
f.block()
```

> **Note**
>
> Flow follows a lazy construction pattern: it won't actually run unless you use the `with` function to open it.

Adding Executors to a Flow

To add elements to your Flow, all you need to do is use the `.add()` method. You can add as many elements as you wish.

The `.add()` method is used to add an Executor to a Flow object. Each `.add()` instance adds a new Executor, and these Executors can be run as a local thread, local process, remote process, inside a Docker container, or even inside a remote Docker container. You can add as many as you need like this:

```
from jina import Flow
flow = Flow().add().add()
```

Defining an Executor via uses

You can use the `uses` parameter to specify the Executor that you are using. The `uses` parameter accepts multiple value types including class names, Docker images, and (inline) YAML. Therefore, you can add an Executor via this:

```
from jina import Flow, Executor
class MyExecutor(Executor):

    ...
f = Flow().add(uses=MyExecutor)
```

Visualizing a Flow

If you want to visualize your Flow, you can do so with the `.plot()` function. This will create a `.svg` file with the visualized Flow. To do this, add the `.plot()` function at the end of your Flow and use the intended title of your `.svg` file:

```
from jina import Flow
f = Flow().add().plot('f.svg')
```

The preceding snippet will produce the following figure with the corresponding Flow:

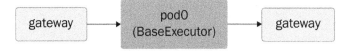

Figure 4.4 – Example of a Flow

> **Note**
>
> In Jupyter Lab/Notebook, the Flow object is rendered automatically without needing to call `.plot()`.

You can also use CRUD methods (index, search, update, delete), which are just sugary syntax forms of post with `on='/index'`, `on='/search'`. These are laid out in the following list:

- `index = partialmethod(post, '/index')`
- `search = partialmethod(post, '/search')`
- `update = partialmethod(post, '/update')`
- `delete = partialmethod(post, '/delete')`

So, taking the previous concepts together, a minimal working example requires the creation of an Executor that extends from the base Executor class and can be used together with your Flow:

```python
from jina import Flow, Document, Executor, requests
class MyExecutor(Executor):
@requests(on='/bar')
  def foo(self, docs, **kwargs):
    print(docs)
f = Flow().add(name='myexec1', uses=MyExecutor)
with f:
f.post(on='/bar', inputs=Document(), on_done=print)
```

That's it! You now have a minimal working example and have covered the basics of Jina. We'll see more advanced uses in the next chapters, but if you've learned the Document, DocumentArray, Executor, and Flow concepts, you are good to go.

Summary

This chapter introduced the main concepts in Jina: Document, DocumentArray, Flow, and Executor. You should now have an overview of what each of those concepts are, why they are important, and how they relate to each other.

Besides understanding the theory of why Document, DocumentArray, Flow, and Executor are important while building your search engine, you should also be able to create a simple Document and assign its corresponding attributes. As you are done with this chapter, you should also be able to create your own Executor and spin up a basic Flow.

You will use all of this knowledge in the next chapter, where you will learn how to integrate these concepts together.

5

Multiple Search Modalities

With the benefits of deep learning and artificial intelligence, we can encode any kind of data into **vectors**. This allows us to create a search system that uses any kind of data as a query and returns any kind of data as a search result.

In this chapter, we will introduce the rising topic of the **multimodal search problem**. You will see different data modalities and how to work with them. You will see how text, images, and audio documents can be transformed into vectors, and how to implement search systems independently of the data modality. You will also see the differences between the concepts of **multimodality** and **cross-modality**.

In this chapter, we're going to cover the following main topics:

- How to represent documents of different data types
- How to encode multimodal documents
- Cross-modal and multimodal searches

By the end of this chapter, you will have a solid understanding of how cross-modal and multimodal searches work, and how easy it is to process data of different modalities in Jina.

Technical requirements

This chapter has the following technical requirements:

- A laptop with a minimum of 4 GB of RAM (8 GB or more is preferred)
- Python installed with 3.7, 3.8, or 3.9 on a Unix-like operating system, such as macOS or Ubuntu

Introducing multimodal documents

Over the last decade, various types of data, such as **texts**, **images**, and **audio**, have been growing rapidly on the internet. Commonly, different types of data are associated with one piece of content. For example, images often also have textual tags and captions to describe the content. Therefore, the content has two modalities: image and text. A movie clip with subtitles has three modalities: image, audio, and text.

Jina is a **data-type-agnostic framework**, letting you work with any type of data and develop cross-modal and multimodal search systems. To better understand what this implies, it makes sense to first show how to represent documents of different data types, and then show how to represent multimodal documents in Jina.

Text document

To represent a textual document in Jina is quite easy. You can do it simply by using the following code:

```
from docarray import Document
doc = Document(text='Hello World.')
```

In some cases, one document can include thousands of words. But a long document with thousands of words is hard to search; some finer granularity would be nice. You can do this by segmenting a long document into smaller *chunks*. For example, let's segment this simple document by using the ! mark:

```
from jina import Document
d = Document(text='नमस्ते दुनिया!你好世界!こんにちは世界!Привет мир!')
d.chunks.extend([Document(text=c) for c in d.text.split('!')])
```

This creates five subdocuments under the original document and stores them under .chunks. To see that more clearly, you can visualize it via d.display(), whose output is shown in the following figure:

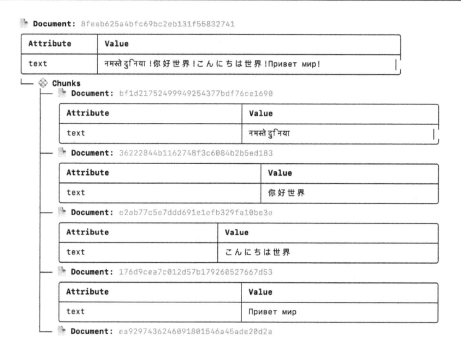

Figure 5.1 – An example of a text document with chunk-level subdocuments

You can also print out each subdocument's text attributes by using the `.texts` sugar attribute:

```
print(d.chunks.texts)
```

This will output the following:

```
['नमस्ते दुनिया', '你好世界', 'こんにちは世界', 'Привет мир', '']
```

That's all you need to know about representing textual documents in Jina!

Image document

Compared to textual data, image data is more universal and easier to comprehend. Image data is often just an **N-dimensional array (ndarray)** – strictly speaking, not just any ndarray, but an ndarray with `ndim=2` or `ndim=3` and `dtype=uint8`. Each element in that ndarray represents a pixel value between 0 and 255 on a certain channel at a certain position. For example, a colored JPG image of 256x300 can be represented as a `[256, 300, 3]` ndarray. You may ask why 3 is in the last dimension. It is because it represents the R, G, and B channels of each pixel. Some images have a different number of channels. For example, a PNG with a transparent background has four channels, where the extra channel represents opacity. A grayscale image has only one channel, which represents the luminance (a measure representing the proportions of black and white).

In Jina, you can load image data by specifying the image URI and then convert it into `.tensor` using the Document API. As an example, we will use the following code to load a PNG apple image (as shown in *Figure 5.2*):

Figure 5.2 – An example PNG image located in the apple.png local file

```
from docarray import Document
d = Document(uri='apple.png')
d.load_uri_to_image_tensor()
print(d.content_type, d.tensor)
```

You will get the following output:

```
tensor [[[255 255 255]
  [255 255 255]
  [255 255 255]

  ...
```

Now the image content is converted into a document's `.tensor` field, which can then be used for further processing. Some help functions can be used to process the image data. You can resize it (that is, downsample/upsample) and normalize it. You can switch the channel axis of `.tensor` to meet certain requirements of some framework, and finally, you can chain all these processing steps together in one line. For example, the image can be normalized and the color axis should be placed first, not last. You can perform such image transformations with the following code:

```
from docarray import Document

d = (
    Document(uri='apple.png')
    .load_uri_to_image_tensor()
    .set_image_tensor_shape(shape=(224, 224))
    .set_image_tensor_normalization()
```

```
        .set_image_tensor_channel_axis(-1, 0)
)

print(d.content_type, d.tensor.shape)
```

You can also dump .tensor back to a PNG image file by using the following:

```
d.save_image_tensor_to_file('apple-proc.png', channel_axis=0)
```

Note that the channel axis is now switched to 0 because of the processing steps we just conducted. Finally, you will get the resulting image shown in *Figure 5.3*:

Figure 5.3 – The resulting image after resizing and normalizing

Audio document

As an important format for storing information, digital audio data can be a soundbite, music, a ringtone, or background noise. It often comes in .wav and .mp3 formats, where the sound waves are digitized by sampling them at discrete intervals. To load a .wav file as a document in Jina, you can simply use the following code:

```
from docarray import Document

d = Document(uri='hello.wav')
d.load_uri_to_audio_tensor()

print(d.content_type, d.tensor.shape)
```

You will see the following output:

```
tensor [-0.00061035 -0.00061035 -0.00082397
...   0.00653076   0.00595093 0.00631714]
```

As shown in the preceding example, the data from the `.wav` file is converted to a one-dimension (mono) ndarray, in which each element is generally expected to lie in the range [-1.0, +1.0] scale. You are by no means restricted to using Jina-native methods for audio processing. Here are some command-line tools, programs, and libraries that you can use for more advanced handling of audio data:

- **FFmpeg** (`https://ffmpeg.org/`): This is a free, open source project for handling multimedia files and streams.

- **Pydub** (`https://github.com/jiaaro/pydub`): This manipulates audio with a simple and easy-to-use high-level interface.

- **Librosa** (`https://librosa.github.io/librosa/`): This is a Python package for music and audio analysis.

Multimodal document

So far, you have learned how to represent different data modalities in Jina. However, in the real world, data often comes in a form that combines multiple modalities, such as video, which typically includes at least *image* and *audio*, as well as *text* in the form of subtitles. Now, it is very interesting to know how to represent multimodal data.

A Jina document can be nested vertically via chunks. It is intuitive to put data of different modalities into subdocuments in chunks. For example, you can create a fashion product document that has two modalities, including a dress image and a product description.

Men Town Round Red Neck T-Shirts

Figure 5.4 – An example of a fashion product document with two modalities

You can do it simply by using the following code:

```
from jina import Document
text_doc = Document(text='Men Town Round Red Neck T-Shirts')
image_doc = Document(uri='tshirt.jpg').load_uri_to_image_
tensor()
fashion_doc = Document(chunks=[text_doc, image_doc])
```

Now, the example fashion product (as shown in *Figure 5.4*) is represented as a Jina document, which has two chunk-level documents representing the product's description and dress image, respectively. You can also use `fashion_doc.display()` to produce the visualization, as shown in *Figure 5.5*:

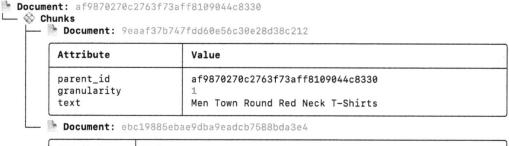

Document: af9870270c2763f73aff8109044c8330
└── ◈ **Chunks**
    ├── **Document:** 9eaaf37b747fdd60e56c30e28d38c212

Attribute	Value
parent_id granularity text	af9870270c2763f73aff8109044c8330 1 Men Town Round Red Neck T-Shirts

    └── **Document:** ebc19885ebae9dba9eadcb7588bda3e4

Attribute	Value
parent_id granularity tensor mime_type uri	af9870270c2763f73aff8109044c8330 1 `<class 'numpy.ndarray'>` in shape (245, 200, 3), dtype: uint8 image/jpeg red_t_shirt.jpg

Figure 5.5 – An example of a fashion product document with two chunk-level documents

> **Important Note**
>
> You may think that different modalities correspond to different kinds of data (images and text in this case). However, this is not accurate. For example, you can do a cross-modal search by searching for images from different perspectives or searching for matching titles for given paragraph text. Therefore, we can consider that a modality is related to a given data distribution from which the data may come.

So far, we have learned how to represent a single piece of text, image, and audio data, as well as representing multimodal data as a Jina document. In the following section, we will show how to get the embedding of each document.

How to encode multimodal documents

After defining the document for different types of data, the next step is to encode the documents into vector embeddings using a model. Formally, embedding was a multi-dimension of a document (often a [1, D] vector), which was designed to contain the content information of a document. With current advances in the performance of all the deep learning methods, even general-purpose models (for example, CNN models trained on ImageNet) can be used to extract meaningful feature vectors. In the following sections, we will show how to encode embedding for documents of different modalities.

Encoding text documents

To convert textual documents into vectors, we can use the pretrained Bert model (https://www.sbert.net/docs/pretrained_models.html) provided by Sentence Transformer (https://www.sbert.net/), as shown in the following example:

```
from docarray import DocumentArray
from sentence_transformers import SentenceTransformer

da = DocumentArray(...)
model = SentenceTransformer('all-MiniLM-L6-v2')
da.embeddings = model.encode(da.texts)

print(da.embeddings.shape)
```

As a result, each document in the input DocumentArray will have an embedding with a 384-dimensional dense vector space after .encode(...) has been completed.

Encoding image documents

For encoding image documents, we can use a pretrained model from Pytorch for embedding. As an example, we will use the **ResNet50** network (https://arxiv.org/abs/1512.03385) for object classification on images provided by **torchvision** (https://pytorch.org/vision/stable/models.html):

```
from docarray import DocumentArray
import torchvision

da = DocumentArray(...)

model = torchvision.models.resnet50(pretrained=True)
```

```
da.embed(model)

print(da.embeddings.shape)
```

In this way, we've successfully encoded an image document into its feature vector representation. The feature vector generated is the output activations of the neural network (a vector of 1,000 components).

> **Important Note**
>
> You might have noticed that in the preceding example, we use `.embed()` for embeddings. Usually, when `DocumentArray` has `.tensors` set, you can use this API for encoding documents. You can specify `.embed(..., device='cuda')` when working with a GPU. The device name identifier depends on the model framework that you are using.

Encoding audio documents

To encode the sound clips into vectors, we chose the **VGGish** model (`https://arxiv.org/abs/1609.09430`) from Google Research. We will use the pretrained model from **torchvggish** (`https://github.com/harritaylor/torchvggish`) to get the feature embeddings for audio data:

```
import torch
from docarray import DocumentArray

model = torch.hub.load('harritaylor/torchvggish', 'vggish')
model.eval()

for d in da:
    d.embedding = model(d.uri)[0]
print(da.embeddings.shape)
```

The returned embeddings for each sound clip are a matrix of the size $K \times 128$, where K is the number of examples in the log mel spectrogram and roughly corresponds to the length of audio in seconds. Therefore, each 4-second audio clip in the chunks is represented by four 128-dimensional vectors.

We have now learned how to encode embeddings for different modalities of documents. In the following section, we will show you how to search for data by using multiple modalities. This can be useful when trying to find data that is not easily represented in a single modality. For example, you might use an image search to find data that is textual in nature.

Cross-modal and multimodal searches

Now that we know how to work with multimodal data, we can describe **cross-modal** and **multimodal** searches. Before that, I would like to first describe the **unimodal** (single-modality) search. In general, unimodal search means processing a single modality of data at both index and query time. For example, in an image search retrieval, the returned search results are also images based on the given image query.

So far, we already know how to encode document content into feature vectors to create embeddings. In the index, each document with the content of an image, text, or audio can be represented as embedding vectors and stored in an indexe. In the query, the query document can also be represented as an embedding, which can then be used to identify similar documents via some similarity scores such as cosine, Euclidean distance, and so on. *Figure 5.6* illustrates the unified matching view of the search problem:

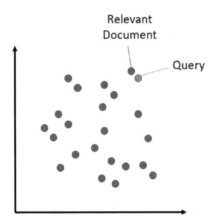

Figure 5.6 – An illustration of the unified matching view for the search problem

More formally, searching can be considered as constructing a matching model that calculates the matching degree between input query documents and documents in the search. With this unified matching view, the matching models for **unimodal**, **multimodal**, and **cross-modal** searches bear even more resemblance to each other in terms of architecture and methodology, as reflected in the techniques: embedding the inputs (queries and documents) as distributed representations, combining neural network components to represent the relations between data of different modalities, and training the model parameters in an end-to-end manner.

Cross-modal search

In a unimodal search, the search is designed to deal with a single data type, making it less flexible and more fragile regarding the input of different data types. Beyond unimodal search, **cross-modal search** aims to take one type of data as the query to retrieve relevant data of another type, such as image-text, video-text, and audio-text cross-modal searches. For example, as shown in *Figure 5.7*, we can devise a text-to-image search system that retrieves images based on short text descriptions as queries:

Figure 5.7 – A cross-modal search system to look for images from captions

More recently, cross-modal search has attracted considerable attention due to the rapid growth of multimodal data. As multimodal data grows, it becomes difficult for users to search for information of interest effectively and efficiently. So far, there have been various search methods proposed for searching multimodal data. However, these search techniques are mostly single modality-based, which converts cross-modal search into keyword-based search. This can be expensive because you need a person to write those keywords, and also, information about multimodal content is not always available. We need to look for another solution! **Cross-modal** search aims to identify relevant data across different modalities. The main challenge in cross-modal search is how to measure the content similarity between different modalities of data. Various methods have been proposed to deal with such a problem. One common way is to generate feature vectors from different modalities in the same latent space, such that newly generated features can be applied in the computation of distance metrics.

To achieve this goal, usually, two very common architectures for deep metric learning (Siamese and triplet networks) can be utilized. They both share the idea that different subnetworks (which may or may not share weights) receive different inputs at the same time (positive and negative pairs for Siamese networks, and positive, negative, and anchor documents for triplets), and try to project their own feature vectors onto a common latent space where the contrastive loss is computed and its error propagated to all the subnetworks.

Positive pairs are pairs of objects (images, text, or any document) that are semantically related and expected to remain close in the projection space. On the other hand, negative pairs are pairs of documents that should be apart.

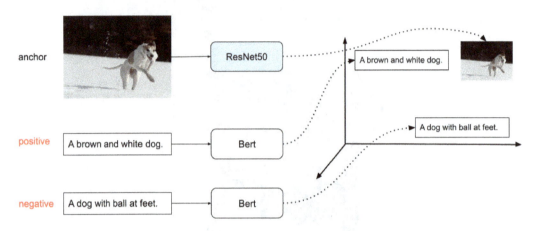

Figure 5.8 – The schema of the deep metric learning process with a triplet network and anchor

As shown in *Figure 5.8*, an example of a cross-modal search between an image and text, the subnetwork used to extract image features is a **ResNet50** architecture with weights pretrained on ImageNet, while for the text embedding, the output of a hidden layer from a pretrained **Bert** model is used. And recently, a new deep metric learning pretrained model, **Contrastive Language-Image Pretraining (CLIP)**, was proposed, which is a neural network trained on a variety of image-text pairs. It is trained to learn visual concepts from natural language with the help of text snippets and image pairs from the internet. It can perform zero-shot learning by encoding text labels and images in the same semantic space and creating a standard embedding for both modalities. With the CLIP-style model, both images and query texts can be mapped into the same latent space, so that they can be compared using a similarity measure.

Multimodal search

Compared to unimodal and cross-modal searches, **multimodal search** aims to enable multimodal data as the query input. The search queries can be composed of a combination of text input, image input, and other modalities of input. It is intuitive to combine different modalities of information for improving search performance. Imagine an e-commerce search scenario that takes two types of query information: an image and a text. For example, if you were searching for pants, the image would be a picture of pants, and the text would be something like "tight" and "blue." In this case, the search query is composed of two modalities (text and image). We can refer to this search scenario as a multimodal search.

To allow for a multimodal search, two approaches are widely used in practice to fuse multiple modalities in the search: early fusion (which fuses features from multiple modalities as query input) and late fusion (which fuses the search results from different modalities at the very end).

Specifically, the early-fusion method fuses the features extracted from data of different modalities. As shown in *Figure 5.9*, features of two different modalities (image and text) resulting from different models are fed into a fusion operator. To combine features simply, we can use the feature concatenation as the fusion operator to produce the feature for multimodal data. Another fusion option involves the projection of different modalities into a common embedding space. And then, we can directly add the features from the data of different modalities.

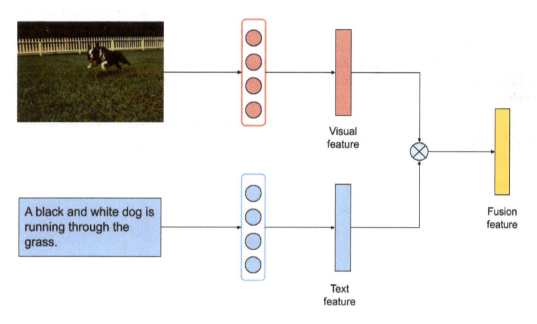

Figure 5.9 – Early fusion, the fusion of multimodal features as the query input

After combining the features, we can use the same method devised for unimodal search to resolve the multimodal search problem. This feature fusion approach suffers from one significant limitation: it would be hard to define the right balance between the importance of various modalities in the context of a user query. To overcome this limitation, an end-to-end neural network can be trained to model the joint multimodal space. However, modeling this joint multimodal space requires a complex training strategy and thoroughly annotated datasets.

In practice, to address the abovementioned shortcomings, we can simply use the late-fusion approach to separate search per modality, and then fuse the search results from different modalities, for example, with a linear combination of the retrieval scores of all modalities per document. While late fusion has been proven to be robust, it has a few issues: appropriate weights of modalities are not a trivial problem, and there is a primary modality issue. For example, in a text-image multimodal search, when the results are assessed by the user based on visual similarity only, the influence of textual scores may worsen the visual quality of the end results.

The main difference between these two search modes is that for cross-modal, there is a direct mapping between a single document and a vector in embedding space, while for multimodal, this does not hold true, since two or more documents might be combined into a single vector.

Summary

This chapter describes the concept of multimodal data, and cross-modal and multimodal search problems. First, we introduced multimodal data and how to represent it in Jina. Then, we learned how to use a deep neural network to get the vector features from data of different modalities. Finally, we introduced cross-modal and multimodal search systems. This unlocks a lot of powerful search patterns and makes it easier to understand how to implement cross-modal and multimodal search applications with Jina.

In the next chapter, we will introduce some basic practical examples to explain how to use Jina to implement search applications.

Part 3: How to Use Jina for Neural Search

In this part, you will use all the knowledge learned so far and you will see step-by-step guides on how to build a search system for different modalities, either for text, images, audio, or cross- and multi-modality. The following chapters are included in this part:

6

Building Practical Examples with Jina

In this chapter, we will build simple real-world applications using Jina's neural search framework. Building on the concepts we have learned in the previous chapters, we will now look at how to use Jina to create valuable applications.

We will learn about the practical aspects of the Jina framework and how you can leverage them to quickly build and deploy sophisticated search solutions. We will walk you through the code base of three different applications built on Jina, and see how the different components that you learned about in the previous chapter work in tandem to create a search application.

We will cover the following three examples in this chapter, which will get you started on the journey of building with Jina:

- The Q/A chatbot
- Fashion image search
- Multimodal search

With this chapter, we aim to get you started by building practical examples to understand the potential of Jina's neural search framework. It is a great stepping stone for venturing into the world of neural search for building state-of-the-art search solutions.

Technical requirements

To follow along with the application code discussed in this chapter, clone the GitHub repository available at `https://github.com/jina-ai/jina/tree/master/jina/helloworld`.

Getting started with the Q/A chatbot

The **Q/A chatbot** is a pre-built example that comes with the Jina installation. To experience the power of Jina firsthand and quickly get started, you can run the Q/A chatbot example directly from the command line without even getting into the code. The Q/A chatbot uses the public Covid Q/A dataset (`https://www.kaggle.com/datasets/xhlulu/covidqa`) from Kaggle, which contains 418 Q/A pairs (`https://www.kaggle.com/xhlulu/covidqa`).

Follow these instructions to set up the development environment and run the Q/A chatbot example:

1. The first step is to install the Jina library from the **Python Package Index** (**PyPI**) along with the required dependencies:

```
pip install "jina[demo]"
```

2. After that, simply type the following command to launch your app:

```
jina hello chatbot
```

After typing this command, you will see the following text on your **command-line interface** (**CLI**):

Figure 6.1 – Q/A chatbot command line

If your screen displays the same text on the command line, it means you have successfully launched the Q/A chatbot example. Now, it's time to open the **user interface** (**UI**) and play with the chatbot.

By default, a simple chat interface will open up, allowing you to chat with the Q/A chatbot. If the page doesn't open up itself, you can open `index.html` by going to `jina/helloworld/chatbot/static`.

You will see the following web page either by default or after opening the `index.html` file:

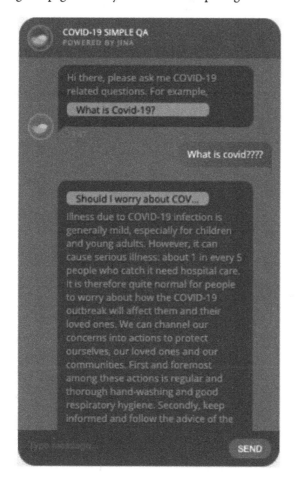

Figure 6.2 – Q/A chatbot interface

You have successfully launched the Q/A chatbot application; it's time to play with it and have some fun. You can ask the chatbot for any Covid-related facts, figures, or queries and see the magic in action!

Navigating through the code

Let's now go through the logic behind the application and see how Jina's framework ties all the components together to produce a functioning Q/A chatbot application.

In order to see the code and understand the different components that work together to bring up this application after installing Jina, go to the chatbot directory by following the `jina/helloworld/chatbot` path. This is the main directory that contains the code for the chatbot example:

```
└── chatbot
    ├── app.py
    ├── my_executors.py
    ├── static/
```

The following are the files that you will see within the chatbot directory:

- `app.py`: This is the main entry point/brain of the application.
- `my_executors.py`: This file is responsible for all the backend processing. It includes the logic behind the application, which we call **executors** in Jina terminology. It hosts multiple executors to transform, encode, and index the data.
- `static`: This folder hosts all the frontend code responsible for rendering the chatbot interface on the web browser that helps you interact with the chatbot application.

We will have a detailed look at the functioning of each of these files in the following subsections.

app.py

The `app.py` file is the entry point of the example application. As soon as you type the `jina hello chatbot` command, the control goes to this file. It's the main entry point for the application and performs all the major tasks of bringing up the application's UI and running the backend code.

The `app.py` file performs the following tasks to ensure that multiple components work in tandem with each other to produce the desired result.

The first thing it does is import the required executors from the `my_executors.py` file using the following code:

```
from my_executors import MyTransformer, MyIndexer
```

Both of these executors are derived from the base `Executor` class of Jina:

- The `MyTransformer` executor is responsible for encoding and transforming the data.
- The `MyIndexer` executor is used for indexing the data.

We will learn about the functioning of both of these executors in detail when we talk about the `my_executors.py` file.

`Flow` allows you to add encoding and indexing in the form of executors, and in the chatbot example, we use the following executors. You can use the following code to create a flow and add these executors to it:

```
from jina import Flow
flow = (
    Flow(cors=True)
    .add(uses=MyTransformer)
    .add(uses=MyIndexer)
    )
```

This is one of the simple flows with just two executors. For a complex flow with many executors, Jina provides the functionality to distinguish each of the executors with distinct names (for example, by using the `name` parameter, you can give your executors some really cool names). It then allows you to visualize the flow to understand how your data flows through different components. Let's visualize this flow by adding a single line to the existing code:

```
from jina import Flow
flow = (
    Flow(cors=True)
    .add(name='MyTransformer', uses=MyTransformer)
    .add(name='MyIndexer', uses=MyIndexer)
    .plot('chatbot_flow.svg')
    )
```

Running the preceding code will generate the following SVG file that visualizes the chatbot flow:

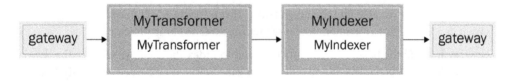

Figure 6.3 – Chatbot flow

> **Note**
> Since we want to call our flow from the browser, it's important to enable Cross-Origin Resource Sharing (`https://developer.mozilla.org/en-US/docs/Web/HTTP/CORS`) within Flow (`cors=True`).

Once we have the flow ready, it's time to dive into the `hello_world` function in the `app.py` file, which brings together everything from different sources and opens a query endpoint (a backend endpoint) for you to query and interact with the chatbot application:

1. The `hello_world` function starts by creating a `workspace` directory to store the indexed data and ensures that the required dependencies are imported.

> **Note**
>
> To run this example, we require two major dependencies/Python libraries: `torch` and `transformers`.

2. Install the dependencies by using the following commands before we move forward with the code:

    *   `pip install torch`

    *   `pip install transformers`

    After installing these dependencies, it's time to continue with the `hello_world` function.

3. The next step is to download the data from Kaggle. For that, we will use the `download_data` function, which basically uses the `urllib` library to fetch and save the data from the given URL.

    The `urllib` module takes `url` and `filename` as the target and downloads the data. You can refer to the following code to see how we set the target:

    ```
    targets = {
            'covid-csv': {
                'url': url_of_your_data,
                'filename': file_name_to_be_fetched,
            }
        }
    ```

    Passing the target variable into the `download_data` function will download the data and save it as a `.csv` file in a random folder within the same working directory.

4. Now we have all the basic components required to index the data, we will use the dataset downloaded in the previous step and index it using the flow that we created previously. Indexing will follow this logic:

    *   It will use the `MyTransformer` executor to encode and transform the data by computing the corresponding embeddings.

    *   It will use the `MyIndexer` executor to index the data via the `/index` endpoint and open the `/search` endpoint to query and interact with the chatbot.

The following is the code that indexes the data and creates a search endpoint to interact with the chatbot:

```
with f:
  f.index(
    DocumentArray.from_csv(
      targets['covid-csv']['filename'],
        field_resolver={'question': 'text'}
    ),
    show_progress=True,)
  url_html_path = 'file://' + os.path.abspath(
    os.path.join(os.path.dirname(
      os.path.realpath(__file__)),'static/index.html'
    )
  )
  try:
    webbrowser.open(url_html_path, new=2)
  except:
    pass
  finally:
    default_logger.info(
      f'You should see a demo page opened in your
      browser,'f'if not, you may open {url_html_path}
      manually'
    )
  if not args.unblock_query_flow:
    f.block()
```

In the preceding code, we open the flow and the dataset with a context manager and send the data in the form of a `'question'`: `'text'` pair to the index endpoint. For this example, we will use the web browser to interact with the chatbot, which requires configuring and serving the flow on a specific port with the HTTP protocol using the `port_expose` parameter, so that the web browser can make requests to the flow. Toward the end, we will use `f.block()` to keep the flow open for search queries and to prevent it from exiting.

my_executors.py

The other key component of the chatbot example is the `my_executors.py` file, which contains the logical elements of the application, also known as **executors**. It consists of two different executors, which we will discuss in detail.

The MyTransformer executor

The `MyTransformer` executor performs the following tasks:

1. It loads the pre-trained sentence transformer model from the `sentence-transformers` library.

2. It takes in the user arguments and sets up the model parameters (such as `model name/path`) and `pooling strategy`, fetches the tokenizer corresponding to the model, and sets up the device to cpu/gpu, depending on the user's preference:

```python
class MyTransformer(Executor):
  """Transformer executor class """

  def __init__(
    self,
    pretrained_model_name_or_path: str =
    'sentence-transformers/paraphrase-mpnet-base-v2',
    pooling_strategy: str = 'mean',
    layer_index: int = -1,
    *args,
    **kwargs,
  ):
  super().__init__(*args, **kwargs)
  self.pretrained_model_name_or_path =
    pretrained_model_name_or_path
  self.pooling_strategy = pooling_strategy
  self.layer_index = layer_index
  self.tokenizer = AutoTokenizer.from_pretrained(
    self.pretrained_model_name_or_path
  )
  self.model = AutoModel.from_pretrained(
    pretrained_model_name_or_path,
      output_hidden_states=True
  )
  self.model.to(torch.device('cpu'))
```

3. After setting up these parameters, it computes the embedding for the textual data and encodes textual data/question-answer as a key-value pair in the form of an embedding map.

4. Encoding is performed through a `sentence-transformers` model (`paraphrase-mpnet-base-v2`, by default). We get the text attributes of documents in batches and then compute embeddings, which we later set as the embedding attribute for each of the documents.

5. The `MyTransformer` executor exposes only one endpoint, encode, which is called whenever we request the flow, either on a query or index. The endpoint creates embeddings for the indexed or query documents so the search endpoint can use similarity scores to determine the closest match for a given query.

Let's look at a simplified version of the `encode` function for the `MyTransformer` executor that we have in the main chatbot application:

```python
@requests
def encode(self, docs: 'DocumentArray', *args, **kwargs):
  with torch.inference_mode():
    if not self.tokenizer.pad_token:
      self.tokenizer.add_special_tokens({'pad_token':
        '[PAD]'})
      self.model.resize_token_embeddings(len(
        self.tokenizer.vocab))
    input_tokens = self.tokenizer(
                docs[:, 'content'],
                padding='longest',
                truncation=True,
                return_tensors='pt',
    )
    input_tokens = {
      k: v.to(torch.device('cpu')) for k,
        v in input_tokens.items()
          }
    outputs = self.model(**input_tokens)
    hidden_states = outputs.hidden_states
    docs.embeddings = self._compute_embedding(
      hidden_states, input_tokens)
```

The MyIndexer executor

The MyIndexer executor performs the following tasks:

1. It uses a document store (SQLite, in our case) that contains all the documents of DocumentArray. The look and feel of DocumentArray with an external store are almost the same as a regular in-memory DocumentArray, but it makes the process more efficient and allows faster retrieval.

2. The executor exposes two endpoints: index and search. The index endpoint is responsible for taking in the documents and indexing them, while the search endpoint is responsible for traversing the indexed DocumentArray to find the relevant match for the user queries.

3. The search endpoint uses the match method (a built-in method associated with DocumentArray), which returns the top closest match for the query documents using cosine similarity.

Let's look at a simplified version of code for the MyIndexer executor that we have in the main chatbot application:

```python
class MyIndexer(Executor):
  """Simple indexer class """
  def __init__(self, **kwargs):
    super().__init__(**kwargs)
    self.table_name = 'qabot_docs'
    self._docs = DocumentArray(
      storage='sqlite',
      config={
        'connection': os.path.join(
          self.workspace, 'indexer.db'),
        'table_name': self.table_name,
      },
    )
  @requests(on='/index')
  def index(self, docs: 'DocumentArray', **kwargs):
    self._docs.extend(docs)
  @requests(on='/search')
  def search(self, docs: 'DocumentArray', **kwargs):
    """Append best matches to each document in docs
    :param docs: documents that are searched
    :param parameters: dictionary of pairs
      (parameter,value)
```

```
    :param kwargs: other keyword arguments

    """
    docs.match(
        self._docs,
        metric='cosine',
        normalization=(1, 0),
        limit=1,
    )
```

These two executors are the building blocks of the chatbot application, and combining them allows us to create an interactive and intelligent chatbot backend. To interact with the chatbot in the web browser via the UI, you can use the HTML template provided in the `static` folder. Running the application by default will open a web page with the chatbot UI; if it doesn't, then you can open the `index.html` file from the `static` folder.

In this section, we looked at the code behind the Q/A chatbot application for the Covid-19 dataset. The application is a form of text-to-text search engine created using Jina's framework. The same logic can be used to create a variety of text search applications depending on your use case.

In the next section, we will explore how to extend the search capabilities for unstructured data types such as images, and see how Jina's neural search makes it easy to build an image-to-image search engine using the fashion image search example.

Understanding fashion image search

Fashion image search is another pre-built example that comes with the Jina installation, which you can run just like the Q/A chatbot example directly from the comfort of your command line without even getting into the code.

The fashion image search example uses the famous *Fashion-MNIST* dataset of Zalando's article images (`https://github.com/zalandoresearch/fashion-mnist`) consisting of 60,000 training examples and 10,000 examples in the test set. Each example is a 28x28 grayscale image, associated with a label from 10 classes just like the original MNIST dataset.

Each training and test set example is assigned one of the following labels:

Label	Description
0	T-shirt/Top
1	Trouser
2	Pullover

Label	Description
3	Dress
4	Coat
5	Sandal
6	Shirt
7	Sneaker
8	Bag
9	Ankle boot

Table 6.1 – Fashion dataset labels and description

In the previous section, we installed the `jina[demo]` library from PyPI, which took care of all the dependencies required to run this example:

1. Let's go to the command line and run the fashion image search example:

```
jina hello fashion
```

2. After typing this command, you will see the following text on your CLI:

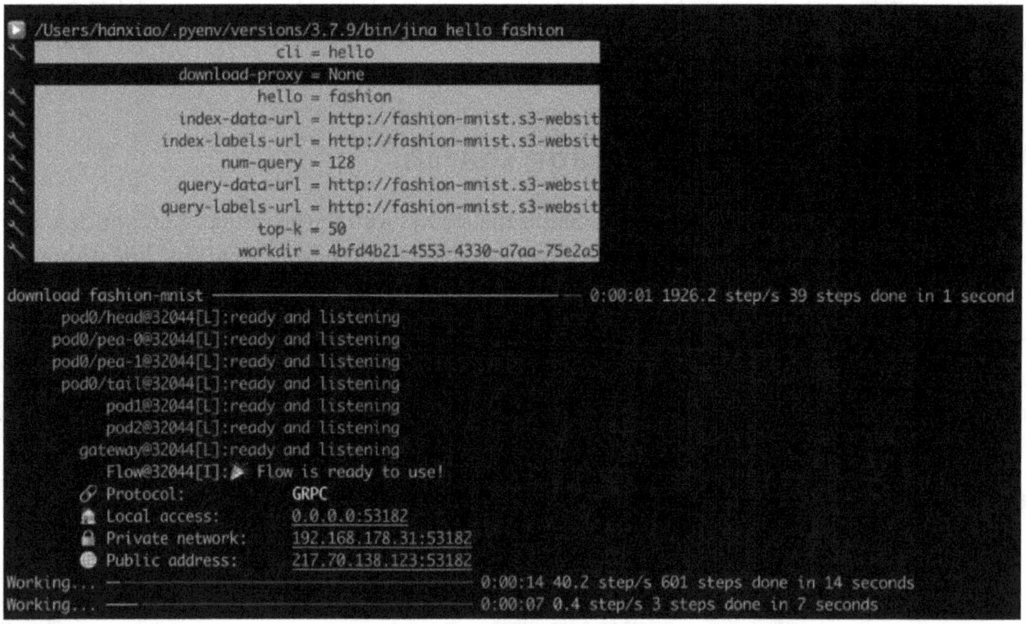

Figure 6.4 – Fashion image search command line

If your screen displays the same text on the command line, it means you have successfully launched the fashion image search example, so now it's time to open the UI and play with the application.

By default, a simple web page will open up with a random sample of images from the test set as queries, along with the retrieved results from the training data. Behind the scenes, Jina downloads the *Fashion-MNIST* dataset and indexes 60,000 training images via the indexing flow. After that, it selects randomly sampled unseen images from the test set as queries and asks Jina to retrieve relevant results.

If the page doesn't open up itself, you can open the demo.html file present at the */demo.html path. You will see the following web page either by default or after opening the downloaded demo.html file manually:

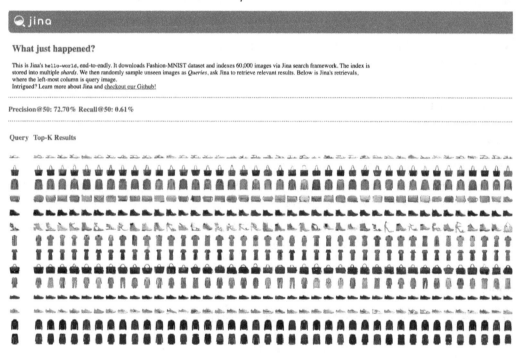

Figure 6.5 – Fashion image search web interface

You can see in the preceding figure how Jina does an amazing job in finding the relevant search results for the image queries selected randomly from the test set.

Navigating through the code

Let's now go through the logic behind the app and see how Jina's framework ties all the components together to create an image search application.

After installing Jina, go to the chatbot directory by following the `jina/helloworld/fashion` path. This is the main directory that contains the code for the fashion image search example:

```
└── fashion
    ├── app.py
    ├── my_executors.py
    ├── helper.py
    ├── demo.html
```

The following are the files that you will see within the fashion directory:

- `app.py`: Similar to the application discussed in the previous section.
- `my_executors.py`: Similar to the application discussed in the previous section.
- `helper.py`: This consists of the supplementary logic functions to modularize the logical code blocks and keep them in a separate file.
- `demo.html`: This hosts all the frontend code responsible for rendering the chatbot interface on the web browser, which helps you interact with the chatbot application.

app.py

The `app.py` file is the entry point of the example application; as soon as you type the `jina hello fashion` command, the control goes to this file. This is the main entry point for the application and performs all the major tasks to bring up the application's frontend and the backend.

The `app.py` file performs the following tasks to ensure that multiple components work in tandem with each other to produce the desired application.

The first thing it does is import the required executors from the `my_executors.py` file using the following code:

```
from my_executors import MyEncoder, MyIndexer
```

All of these executors are derived from the base `Executor` class of Jina:

- `MyEncoder` is responsible for transforming and encoding the data.
- `MyIndexer` is used for indexing the data; after indexing, it hosts a `/search` endpoint for querying the data.

We will learn about the functioning of all these executors in detail when we talk about the `my_executors.py` file. The flow for this example consists of the aforementioned executors.

You can use the following code to create and visualize the flow:

```
from jina import Flow
flow = (
    Flow()
    .add(name='MyEncoder', uses=MyEncoder, replicas=2)
    .add(name='MyIndexer', uses=MyIndexer)
    .plot('fashion_image_flow.svg')
    )
```

Running the code will generate the following flow diagram, which shows how the data moves through different components of the applications:

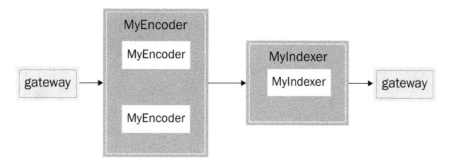

Figure 6.6 – Fashion image search flow

In the preceding code, the `replicas` parameter is set to 2 for the `MyEncoder` executor to divide the input data stream into two different executors for faster processing and encoding.

Once we have the flow ready, it's time to dive into the `hello_world` function in the `app.py` files, which brings together everything from different sources. The `hello_world` function performs the following tasks:

1. It creates a `workspace` directory in the root folder to store the indexed data.

2. It creates a `targets` dictionary to associate the URL of the data with the local filenames where the data will be saved. It saves the training data under the `index` and `index-label` files, and the test data under the `query` and `query-label` files:

```
targets = {
        'index-labels': {
            'url': args.index_labels_url,
            'filename': os.path.join(args.workdir,
            'index-labels'),
```

```
        },
        'query-labels': {
            'url': args.query_labels_url,
            'filename': os.path.join(args.workdir,
            'query-labels'),
        },
        'index': {
            'url': args.index_data_url,
            'filename': os.path.join(args.workdir,
        'index-original'),
    },
        'query': {
            'url': args.query_data_url,
            'filename': os.path.join(args.workdir,
            'query-original'),},
    }
```

3. After that, it passes the targets variable to the download_data function and downloads the *Fashion-MNIST* dataset. The download_data function uses the urllib package to download the data from the given URL and iterate through the dictionary to save the data and the labels for the training and the test set.

4. It creates the flow and adds the MyEncoder and MyIndexer executors.

5. It opens the flow with the context manager and uses the indexing flow to index the data by creating the embeddings for all the images in the training data.

6. It then includes the ground truth (labels) along with the query images, which allows us to evaluate the performance of the model.

7. After indexing the data, it calls the search function, which randomly samples 128 unseen images as queries and returns the top 50 similar images for each of the query images.

8. Finally, we use the write_html function to render the frontend in the web browser using the demo.html file:

```
with f:
    f.index(index_generator(num_docs=targets['index']
        ['data'].shape[0], target=targets),
        show_progress=True,
        )
    groundtruths = get_groundtruths(targets)
```

```
evaluate_print_callback = partial(print_result,
   groundtruths)
evaluate_print_callback.__name__ =
   'evaluate_print_callback'
f.post(
   '/search,
   query_generator(num_docs=args.num_query,
      target=targets),
   shuffle=True,
   on_done= evaluate_print_callback,
   parameters={'top_k': args.top_k},
   show_progress=True,
   )
#write result to html
write_html(os.path.join(args.workdir, 'demo.html'))
```

my_executors.py

The other key component of the fashion image search example is the my_executors.py file. It consists of three different executors that work together in the flow to create an end-to-end application experience.

The MyEncoder executor

The MyEncoder executor performs the following tasks:

1. It is used in both indexing and the querying flow. It is fed with the index and query data yielded from the respective generator functions. It uses **singular value decomposition (SVD)** to encode the incoming data.

2. In the constructor, it creates a random matrix of shape (784, 64) and applies SVD to get oth_mat.

3. In the encode function, it fetches the content from the docs array (DocumentArray in Jina), stacks images together, extracts the single-channel content, and reshapes images to make it ready to fetch the embeddings.

4. In the next step, we use the content matrix along with oth_mat (the result of SVD) to get the embeddings.

5. It then associates each document tensor with the respective embeddings and converts the tensor into a **uniform resource identifier (URI)** (a long string that is an equivalent representation of an image) for standardized representation and then it pops the tensor.

6. It repeats the same process for all the images in the loop to encode the entire dataset:

```
class MyEncoder(Executor):
    """
    Encode data using SVD decomposition
    """
    def __init__(self, **kwargs):
        super().__init__(**kwargs)
        np.random.seed(1337)
        # generate a random orthogonal matrix
        H = np.random.rand(784, 64)
        u, s, vh = np.linalg.svd(H, full_matrices=False)
        self.oth_mat = u @ vh
    @requests
    def encode(self, docs: 'DocumentArray', **kwargs):
        """Encode the data using an SVD decomposition
        :param docs: input documents to update with an
          embedding
        :param kwargs: other keyword arguments
        """
        # reduce dimension to 50 by random orthogonal
        # projection
        content = np.stack(docs.get_attributes('content'))
        content = content[:, :, :, 0].reshape(-1, 784)
        embeds = (content / 255) @ self.oth_mat
        for doc, embed, cont in zip(docs, embeds,
          content):
            doc.embedding = embed
            doc.content = cont
            doc.convert_image_tensor_to_uri()
            doc.pop('tensor')
```

The MyIndexer executor

The `MyIndexer` executor performs the following tasks:

1. Its constructor creates a `workspace` directory to store the indexed data.

2. It hosts an `index` endpoint, which takes in the documents as input and structures them into the `workspace` folder.

3. It also hosts the `search` endpoint, which gives out the best matches for a given query. It takes in the document and `top-k` as a parameter and performs a cosine similarity match to find the `top-k` results:

```python
class MyIndexer(Executor):
  """

  Executor with basic exact search using cosine
  distance
  """

  def __init__(self, **kwargs):
    super().__init__(**kwargs)
    if os.path.exists(self.workspace + '/indexer'):
      self._docs = DocumentArray.load(self.workspace +
      '/indexer')
    else:
      self._docs = DocumentArray()


  @requests(on='/index')
  def index(self, docs: 'DocumentArray', **kwargs):
    """Extend self._docs
    :param docs: DocumentArray containing Documents
    :param kwargs: other keyword arguments
    """
    self._docs.extend(docs)
  @requests(on=['/search', '/eval'])
  def search(self, docs: 'DocumentArray',
    parameters: Dict, **kwargs):
    """Append best matches to each document in docs
    :param docs: documents that are searched
    :param parameters: dictionary of pairs
      (parameter,value)
    :param kwargs: other keyword arguments
    """
    docs.match(
      self._docs,
      metric='cosine',
      normalization=(1, 0),
```

```
        limit=int(parameters['top_k']),
    )
def close(self):
    """

    Stores the DocumentArray to disk
    """

    self._docs.save(self.workspace + '/indexer')
```

helper.py

The `helper.py` file provides the helper functions to support the logical elements in the `app.py` file. It implements key functions such as `index_generator` and `query_generator`, which we use in the `app.py` file to index and query the data. Let's go through both of these functions and understand what they do.

index_generator()

This function generates the index tag for the training data using the following steps:

1. This generator will iterate over all 60,000 documents (images) and process each one individually to make them index-ready.

2. It fetches the 28x28 images from the dictionary and inverts them to make them suitable to be displayed on the web browser.

3. It converts the black and white image into an RGB image and then converts the image into Jina's internal data type, `Document`.

4. It then associates a tag ID with the document and yields it as the index data.

    The following is the code for the `index_generator()` function:

    ```
    def index_generator(num_docs: int, target: dict):
        """

        Generate the index data.
        :param num_docs: Number of documents to be indexed.
        :param target: Dictionary which stores the data
          paths
        :yields: index data
        """

        for internal_doc_id in range(num_docs):
          # x_blackwhite.shape is (28,28)
          x_blackwhite=
    ```

```
            255-target['index']['data'][internal_doc_id]
        # x_color.shape is (28,28,3)
        x_color = np.stack((x_blackwhite,) * 3, axis=-1)
        d = Document(content=x_color)
        d.tags['id'] = internal_doc_id
        yield d
```

query_generator()

This is similar to the index_generator function and follows the same logic to generate the query data with some modifications. It fetches a random number of documents (based on the value of the num_docs parameter) from the dataset to generate the query data. The following is the code for the query_generator() function:

```
def query_generator(num_docs: int, target: dict):
    """

    Generate the query data.
    :param num_docs: Number of documents to be queried
    :param target: Dictionary which stores the data paths
    :yields: query data
    """
    for _ in range(num_docs):
        num_data = len(target['query-labels']['data'])
        idx = random.randint(0, num_data - 1)
        # x_blackwhite.shape is (28,28)
        x_blackwhite = 255 - target['query']['data'][idx]
        # x_color.shape is (28,28,3)
        x_color = np.stack((x_blackwhite,) * 3, axis=-1)
        d = Document(
            content=x_color,
            tags={
            'id': -1,
            'query_label': float(target['query-labels']
              ['data'][idx][0]),
            },
        )
        yield d
```

demo.html

To view the query results in the web browser, the application uses the `demo.html` file to render the frontend. By default, running the application will open a web page with the query images along with the search results; if it doesn't, then you can open the `demo.html` file, which will be available in the random folder generated at the start.

In this section, we saw how Jina's framework makes it really efficient to build search applications for image data types by leveraging state-of-the-art deep learning models. The same functionality will be extended to other data types, such as audio, video, and even 3D mesh, which you will learn about in *Chapter 7, Exploring Advanced Use Cases of Jina.*

Next, we will look at how to combine two data types to create a multimodal search that can easily elevate the search experience for your product or platform. We will dive into the multimodal search example, which uses the *people-image* dataset consisting of *image-caption* pairs to build a search application that lets you query using both the image and the text.

Working with multimodal search

Multimodal search is another pre-built example that comes with the Jina installation, which you can run directly from the comfort of your command line without even getting into the code.

This example uses Kaggle's public people image dataset (`https://www.kaggle.com/ahmadahmadzada/images2000`), which consists of 2,000 image-caption pairs. The data type here is a multimodal document containing multiple data types such as a PDF document that contains text and images together. Jina lets you build the search for multimodal data types with the same ease and comfort:

1. To run this example from the command line, you need to install the following dependencies:

    *   `pip install transformers`

    *   `pip install torch`

    *   `pip install torchvision`

    *   `pip install "jina[demo]"`

2. Once all the dependencies are installed, simply type the following command to launch the application:

    ```
    jina hello multimodal
    ```

3.   After typing this command, you will see the following text on your CLI:

```
/Users/hanxiao/.pyenv/versions/3.7.9/bin/jina hello multimodal
                             cli = hello
                  download-proxy = None
                           hello = multimodal
                  index-data-url = https://static.jina.ai/multimo
                     port-expose = 8080
              unblock-query-flow = False
                         workdir = b7518349-5b91-4255-be7d-1d4857

download zip data ─────────────────────────────── 0:00:16 676.4 step/s    112
       segment@32747[L]:ready and listening
     craftText@32747[L]:ready and listening
    encodeText@32747[L]:ready and listening
   textIndexer@32747[L]:ready and listening
    craftImage@32747[L]:ready and listening
   encodeImage@32747[L]:ready and listening
   imageIndexer@32747[L]:ready and listening
keyValueIndexer@32747[L]:ready and listening
       joinAll@32747[L]:ready and listening
       gateway@32747[L]:ready and listening
          Flow@32747[I]:  Flow is ready to use!
        Protocol:          GRPC
        Local access:      0.0.0.0:54775
        Private network:   192.168.178.31:54775
        Public address:    217.70.138.123:54775
Working... ────────────────────────────── 0:00:22 0.8 step/s    ▉
```

Figure 6.7 – Multimodal search command line

If your screen displays the same text on the command line, it means you have successfully launched the Jina multimodal example; now, it's time to open the UI and play with the application.

By default, a UI with a query and results section will open up, allowing you to query with text and image and get the results in the same form. If the page doesn't open up itself, you can open the index. html file by going to jina/helloworld/multimodal/static.

You will see the following web page either by default or after opening the `index.html` file:

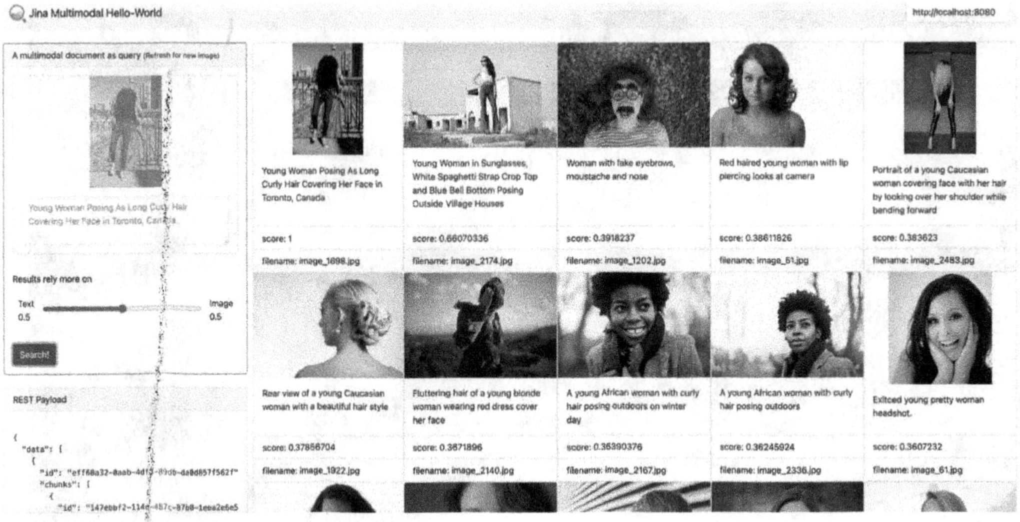

Figure 6.8 – Multimodal search interface

You have successfully launched the multimodal example application; it's now time to play with it and have some fun.

Navigating through the code

Let's now go through the logic behind the app and see how Jina's framework ties all the components together to produce a functioning multimodal search application.

Once you install Jina, go to the chatbot directory by following the `jina/helloworld/multimodal` path. This is the main directory and contains the code for the multimodal search example:

```
└── multimodal
    ├── app.py
    ├── my_executors.py
    ├── flow_index.yml
    ├── flow_search.yml
    ├── static/
```

The following are the files that you will see within the multimodal directory. We will go through the functioning of each of them in detail:

- `app.py`: Similar to the previous applications.
- `my_executors.py`: Similar to the previous applications.

- The `static` folder: This hosts all the frontend code responsible for rendering the UI on the web browser, which helps you interact with the application.

- `flow_index.yml`: This contains the YAML code for the indexing flow, which is run when we index the data for the first time.

- `flow_search.yml`: This contains the YAML code for the search flow, which runs every time we send any query to the application.

This application uses the MobileNet and MPNet models to index the image-caption pairs. The indexing process takes about 3 minutes on the CPU. Then, it opens a web page where you can query the multimodal documents. We have also prepared a YouTube video (`https://youtu.be/B_nH8GCmBfc`) to walk you through this demo.

app.py

When you type the `jina hello multimodal` command, the control of the application goes to the `app.py` file. The `app.py` file performs the following tasks to ensure that all the components of the multimodal search application work in tandem with each other to produce the desired result.

The first thing it does is import the required libraries. After that, the control goes to the `hello_world()` function, which hosts the main logic of the script. The `hello_world()` function creates a random directory using the `mkdir` command to store the artifacts, such as the downloaded data. Then, it checks to ensure that all the required Python libraries are installed and imported.

> **Note**
>
> To run this example, we require three major dependencies/Python libraries: `torch`, `transformers`, and `torchvision`.

Following are the steps to understand the functioning of `app.py` file:

1. Please check that all the aforementioned dependencies are installed correctly in your Python environment.

2. After checking that these dependencies are correctly installed, the `hello_world()` function calls the `download_data()` function to fetch and download the data from Kaggle. The `download_data()` function uses the `urllib` package to fetch and save the data from the given URL. `urllib` takes the URL and filename as the targets and downloads the data. You can refer to the following code to see how we set the targets:

```
targets = {
        'people-img: {
            'url': url_of_the_data,
            'filename': file_name_to_be_fetched,
```

```
        }
    }
```

Passing the `targets` variable into the `download_data()` function will download the data and save it in the random folder created at the beginning of the `hello_world` function. It then loads the indexing flow from the YAML file and passes the image metadata to the flow:

```
# Indexing Flow
f = Flow.load_config('flow-index.yml')


with f, open(f'{args.workdir}/people-img/meta.csv',
newline='') as fp:
  f.index(inputs=DocumentArray.from_csv(fp),
    request_size=10, show_progress=True)
  f.post(on='/dump', target_executor='textIndexer')
  f.post(on='/dump', target_executor='imageIndexer')
  f.post(on='/dump',
    target_executor='keyValueIndexer')
```

3. Similarly, it then loads the search flow from the YAML file and sets it to fetch the input queries from the HTML frontend:

```
# Search Flow
f = Flow.load_config('flow-search.yml')


# switch to HTTP gateway
f.protocol = 'http'
f.port_expose = args.port_expose
url_html_path = 'file://' + os.path.abspath(
            os.path.join(cur_dir,
            'static/index.html'))
with f:
  try:
        webbrowser.open(url_html_path, new=2)
  except:
    pass  # intentional pass
  finally:
        default_logger.info(
    f'You should see a demo page opened in your
```

```
        browser,'f'if not, you may open {url_html_path}
        manually'
        )
    if not args.unblock_query_flow:
        f.block()
```

In both of the preceding code snippets, we open the flow with a context manager. For this example, we will use the web browser to interact with the application. It requires configuring and serving the flow on a specific port with the HTTP protocol using the port_expose parameter. Toward the end, we use the f.block() method to keep the flow open for search queries and to prevent it from exiting.

my_executors.py

If app.py is the brain of this example, then the my_executors.py file contains the neurons in the form of executors that power the core logic.

The multimodal example contains two modalities of data: image and text, which are stored in the document tags and uri attributes, respectively. To process these two modalities of data, at index time, we need to preprocess, encode, and index them separately using the following executors.

The Segmenter executor

The Segmenter executor takes the documents as the input and splits them into two chunks: image chunk and text chunk. The text chunk will contain the plain text data and the image chunk (which we call chunk_uri in the code) contains the URI of the image. Then, we add them both to the document chunks and send them further to the pre-processing stage, as shown here:

```
class Segmenter(Executor):
    @requests
    def segment(self, docs: DocumentArray, **kwargs):
        for doc in docs:
            text = doc.tags['caption']
            uri={os.environ["HW_WORKDIR"]}/
              people-img/{doc.tags["image"]}'
            chunk_text = Document(text=text,
              mime_type='text/plain')
            chunk_uri = Document(uri=uri,
              mime_type='image/jpeg')
            doc.chunks = [chunk_text, chunk_uri]
            doc.uri = uri
            doc.convert_uri_to_datauri()
```

The TextCrafter executor

For the preprocessing of the text chunk, we use the `TextCrafter` executor, which takes the text chunk as the input and returns a flattened traversable sequence of all the documents, as shown here:

```
class TextCrafter(Executor):
    def __init__(self, *args, **kwargs):
        super().__init__(*args, **kwargs)
    @requests()
    def filter(self, docs: DocumentArray, **kwargs):
        filtered_docs = DocumentArray(
            d for d in docs.traverse_flat(['c']) if
                d.mime_type == 'text/plain'
        )
        return filtered_docs
```

The ImageCrafter executor

Similarly, for the preprocessing of the image chunk, we use the `ImageCrafter` executor, which takes the image chunk as the input and returns a flattened traversable sequence of all the documents:

```
class ImageCrafter(Executor):
    @requests(on=['/index', '/search'])
    def craft(self, docs: DocumentArray, **kwargs):
        filtered_docs = DocumentArray(
            d for d in docs.traverse_flat(['c']) if
                d.mime_type == 'image/jpeg'
        )
        target_size = 224
        for doc in filtered_docs:
            doc.convert_uri_to_image_blob()
            doc.set_image_blob_shape(shape=(target_size,
                target_size))
            doc.set_image_blob_channel_axis(-1, 0)
        return filtered_docs
```

The TextEncoder executor

After the preprocessing step, the preprocessed data of the text chunk goes to the `TextEncoder` executor as the input and produces the text embedding as the output. We persist the result in the form of embeddings using the `DocVectorIndexer` executor. Let's look at the functioning of `TextEncoder` by starting with the code of its constructor:

```python
class TextEncoder(Executor):
    """Transformer executor class"""
    def __init__(
            self,
            pretrained_model_name_or_path: str=
            'sentence-transformers/paraphrase-mpnet-base-v2',
            pooling_strategy: str = 'mean',
            layer_index: int = -1,
            *args,
            **kwargs,
    ):
            super().__init__(*args, **kwargs)
            self.pretrained_model_name_or_path =
                pretrained_model_name_or_path
            self.pooling_strategy = pooling_strategy
            self.layer_index = layer_index
            self.tokenizer = AutoTokenizer.from_pretrained(
                self.pretrained_model_name_or_path
            )
            self.model = AutoModel.from_pretrained(
                self.pretrained_model_name_or_path,
                output_hidden_states=True
            )
            self.model.to(torch.device('cpu'))
```

To compute the embeddings, it uses the pre-trained `sentence-transformers/paraphrase-mpnet-base-v2` model with the `'mean'` pooling strategy. Let's look at the code for the `compute_embedding()` function:

```python
def _compute_embedding(self, hidden_states: 'torch.Tensor',
  input_tokens:   Dict):
    fill_vals = {'cls': 0.0,'mean': 0.0,'max': -np.inf,'min':
      np.inf}
        fill_val = torch.tensor(
            fill_vals[self.pooling_strategy],
```

```
            device=torch.device('cpu')
        )
    layer = hidden_states[self.layer_index]
        attn_mask =
          input_tokens['attention_mask']
          .unsqueeze(-1).expand_as(layer)
        layer = torch.where(attn_mask.bool(), layer,
          fill_val)
        embeddings = layer.sum(dim=1) / attn_mask.sum(dim=1)
        return embeddings.cpu().numpy()
```

It then uses the encode() function to store the embeddings in the doc.embeddings attribute of the document:

```
@requests
def encode(self, docs: 'DocumentArray', **kwargs):
  with torch.inference_mode():
        if not self.tokenizer.pad_token:
            self.tokenizer.add_special_tokens({
                'pad_token': '[PAD]'})
      self.model.resize_token_embeddings(len(
        self. tokenizer.vocab))
    input_tokens = self.tokenizer(
      docs.get_attributes('content'),
      padding='longest',
      truncation=True,
      return_tensors='pt',
        )
        input_tokens = {
    k: v.to(torch.device('cpu')) for k, v in
      input_tokens.items()
        }
        outputs = self.model(**input_tokens)
        hidden_states = outputs.hidden_states
        docs.embeddings = self._compute_embedding(
          hidden_states, input_tokens)
```

The ImageEncoder executor

Similarly, the preprocessed data of the image chunk goes to the `ImageEncoder` executor as the input and produces the embedding as the output. We persist the result in the form of embeddings using the `DocVectorIndexer` executor. Let's look at the functioning of `ImageEncoder` by going through the code:

```python
class ImageEncoder(Executor):
  def __init__(
      self,
    model_name: str = 'mobilenet_v2',
    pool_strategy: str = 'mean',
    channel_axis: int = -1, *args, **kwargs,
  ):
    super().__init__(*args, **kwargs)
    self.channel_axis = channel_axis
    self.model_name = model_name
    self.pool_strategy = pool_strategy
    self.pool_fn = getattr(np, self.pool_strategy)
        model = getattr(models,
          self.model_name)(pretrained=True)
    self.model = model.features.eval()
    self.model.to(torch.device('cpu'))
```

It uses the pre-trained `mobilenet -v2` model to generate the embeddings. To preprocess the images, it uses the `'mean'` pooling strategy to compute the average value of all the pixels in the image to compute the embeddings:

```python
def _get_features(self, content):
  return self.model(content)


def _get_pooling(self, feature_map: 'np.ndarray') -> 'np.
ndarray':
  if feature_map.ndim == 2 or self.pool_strategy is None:
    return feature_map
  return self.pool_fn(feature_map, axis=(2, 3))
@requests
def encode(self, docs: DocumentArray, **kwargs):
  with torch.inference_mode():
```

```
_input = torch.from_numpy(docs.blobs.astype('float32'))
    _features = self._get_features(_input).detach()
    _features = _features.numpy()
    _features = self._get_pooling(_features)
docs.embeddings = _features
```

Toward the end, the encode function stores the embeddings in the doc.embeddings attribute of the document.

The DocVectorIndexer executor

Now, let's look at the DocVectorIndexer executor, which persists the encoding from both TextEncoder and ImageEncoder to index them. Here, we have two different modalities of data (text and image), so we need to store the indexed results separately in two different files. The DocVectorIndexer executor takes care of that. It stores the indexed text embeddings into the text.json file and the image embeddings into the image.json file, which we will use in the flow_index.yml file as index_file_name. Let's look at the code for DocVectorIndexer to understand its functioning in detail:

```
class DocVectorIndexer(Executor):
  def __init__(self, index_file_name: str, **kwargs):
      super().__init__(**kwargs)
    self._index_file_name = index_file_name
    if os.path.exists(self.workspace +
      f'/{index_file_name}'):
      self._docs = DocumentArray.load(
        self.workspace + f'/{index_file_name}')
    else:
      self._docs = DocumentArray()
  @requests(on='/index')
  def index(self, docs: 'DocumentArray', **kwargs):
    self._docs.extend(docs)
  @requests(on='/search')
  def search(self, docs: 'DocumentArray', parameters: Dict,
    **kwargs):
    docs.match(
      self._docs,
      metric='cosine',
            normalization=(1, 0),
```

```
                    limit=int(parameters['top_k']),
    )
  @requests(on='/dump')
  def dump(self, **kwargs):
    self._docs.save(self.workspace +
      f'/{self._index_file_name}')
  def close(self):
    """

    Stores the DocumentArray to disk
    """

    self.dump()
    super().close()
```

It uses `DocumentArray` to store all the documents directly on the disk because we have a large number of documents. It hosts two different endpoints to index the data and open the `'search'` flow. It uses the cosine similarity score to find the relevant documents.

The KeyValueIndexer executor

Apart from `DocVectorIndexer` to persist embeddings, we also create a `KeyValueIndexer` executor to help the chunks (text chunk and image chunk) to find their parent/root document. Let's look at the code to understand its functionality in detail:

```
class KeyValueIndexer(Executor):
  def __init__(self, *args, **kwargs):
    super().__init__(*args, **kwargs)
    if os.path.exists(self.workspace + '/kv-idx'):
      self._docs = DocumentArray.load(self.workspace +
            '/kv-idx')
    else:
      self._docs = DocumentArray()
  @requests(on='/index')
  def index(self, docs: DocumentArray, **kwargs):
    self._docs.extend(docs)
  @requests(on='/search')
  def query(self, docs: DocumentArray, **kwargs):
    for doc in docs:
            for match in doc.matches:
        extracted_doc = self._docs[match.parent_id]
```

```
        extracted_doc.scores = match.scores
        new_matches.append(extracted_doc)
    doc.matches = new_matches
@requests(on='/dump')
def dump(self, **kwargs):
  self._docs.save(self.workspace +
    f'/{self._index_file_name}')


def close(self):
    """

    Stores the DocumentArray to disk
    """
    self.dump()
    super().close()
```

It uses `DocumentArray` just like `DocVectorIndexer` to store all the documents directly on the disk.

It hosts two different endpoints to index the data and open the search flow. In the search logic, given a document, it loops through the tree to find its root/parent document.

The WeightedRanker executor

Toward the end, when both the chunks find their parents, we aggregate the score using the `WeightedRanker` executor to produce the final output.

Let's look at the code to understand its functionality in detail:

1. It opens a search endpoint to combine the results from both the text and image chunks to calculate the final similarity score, which we will use to determine the results:

```
class WeightedRanker(Executor):
  @requests(on='/search')
  def rank(
    self, docs_matrix: List['DocumentArray'],
    parameters: Dict, **kwargs
  ) -> 'DocumentArray':
    """

    :param docs_matrix: list of :class:`DocumentArray`
      on multiple      requests to get bubbled up
```

```
        matches.
    :param parameters: the parameters passed into the
        ranker, in     this case stores
            :param kwargs: not used (kept to maintain
                interface)
    """
    result_da = DocumentArray()
    for d_mod1, d_mod2 in zip(*docs_matrix):
                final_matches = {}  # type: Dict[str,
                    Document]
```

2. You can assign the `weight` parameter beforehand to determine which modality (between text and image) will contribute more toward calculating the final relevance score. If you set the weight of the text chunk as 2 and the image chunk as 1, then the text chunk will contribute a higher score to the final relevance.

3. The final similarity score is calculated by summing up cosine similarity * weight for both the modalities and then sorting them in descending order:

```
    for m in d_mod1.matches:
        relevance_score = m.scores['cosine'].value *
            d_mod1.weight
        m.scores['relevance'].value = relevance_score
        final_matches[m.parent_id] = Document(
            m, copy=True)
    for m in d_mod2.matches:
        if m.parent_id in final_matches:
            final_matches[m.parent_id].scores[
                'relevance'
            ].value = final_matches[m.parent_id].
    scores['relevance']
            .value + (
                m.scores['cosine'].value * d_mod2.weight
            )
        else:
            m.scores['relevance'].value = (
                m.scores['cosine'].value * d_mod2.weight
            )
                final_matches[m.parent_id] = Document(m,
```

```
                    copy=True)
    da = DocumentArray(list(final_matches.values()))
    da.sorted(da, key=lambda ma:
      ma.scores['relevance'].value, reverse=True)
    d = Document(matches=da[: int(parameters['top_k'])])
    result_da.append(d)
return result_da
```

We have looked at how the executors work together to produce the results. Let's now look at how these executors are arranged and utilized in the index and search flow.

flow_index.yml

As you already know, Jina provides two ways to create and work with the flows. The first is by using native Python, and the second is by using a YAML file to create a flow and call it in the main app. py file. Now, we will look at how the flow_index.yml file is created by leveraging the individual executor components that we discussed in the previous section.

The flow_index.yml file uses different executors that we have defined in the my_executors. py file and arranges them to produce the indexing flow. Let's go through the YAML code to understand it in detail:

1. It starts with the Segmenter executor, which segments the document into text and image chunks:

    ```
    jtype: Flow
    version: '1'
    executors:
      - name: segment
        uses:
          jtype: Segmenter
          metas:
            workspace: ${{ ENV.HW_WORKDIR }}
            py_modules:
              - ${{ ENV.PY_MODULE }}
    ```

2. After that, we have two different pipelines, one for text and the other for the image. The text pipeline preprocesses the data using the TextCrafter executor, encodes it using the TextEncoder executor, and then indexes it using DocVectorIndexer:

    ```
      - name: craftText
        uses:
    ```

```
        jtype: TextCrafter
        metas:
          py_modules:
            - ${{ ENV.PY_MODULE }}
  - name: encodeText
    uses:
        jtype: TextEncoder
        metas:
          py_modules:
            - ${{ ENV.PY_MODULE }}
  - name: textIndexer
    uses:
        jtype: DocVectorIndexer
        with:
          index_file_name: "text.json"
        metas:
          workspace: ${{ ENV.HW_WORKDIR }}
          py_modules:
            - ${{ ENV.PY_MODULE }}
```

3. The image pipeline preprocesses the data using the `ImageCrafter` executor, encodes it using the `ImageEncoder` executor, and then indexes it using `DocVectorIndexer`:

```
  - name: craftImage
    uses:
        jtype: ImageCrafter
        metas:
          workspace: ${{ ENV.HW_WORKDIR }}
          py_modules:
            - ${{ ENV.PY_MODULE }}
    needs: segment
  - name: encodeImage
    uses:
        jtype: ImageEncoder
        metas:
          py_modules:
            - ${{ ENV.PY_MODULE }}
```

```
  - name: imageIndexer
    uses:
      jtype: DocVectorIndexer
      with:
        index_file_name: "image.json"
      metas:
        workspace: ${{ ENV.HW_WORKDIR }}
        py_modules:
          - ${{ ENV.PY_MODULE }}
```

4. After indexing the text and image to the respective text.json and image.json files, we join both the indexers with KeyValueIndexer to link them together:

```
  - name: keyValueIndexer
    uses:
      jtype: KeyValueIndexer
      metas:
        workspace: ${{ ENV.HW_WORKDIR }}
        py_modules:
          - ${{ ENV.PY_MODULE }}
    needs: segment
  - name: joinAll
    needs: [textIndexer, imageIndexer,
      keyValueIndexer]
```

flow_search.yml

Similar to the flow_index.yml file, we also have a flow_search.yml file, which defines the search/query flow for the multimodal example application. Let's look at the YAML code to understand its functionality in detail:

1. It gets the input in the form of text and images and treats them both differently using a pipeline of executors. For the text input, it uses the TextCrafter executor to preprocess the data, followed by the TextEncoder executor to encode the textual data, and finally, indexes it using DocVectorIndexer:

```
jtype: Flow
version: '1'
with:
  cors: True
```

```
executors:
  - name: craftText
    uses:
      jtype: TextCrafter
      metas:
        py_modules:
          - ${{ ENV.PY_MODULE }}
  - name: encodeText
    uses:
      jtype: TextEncoder
      metas:
        py_modules:
          - ${{ ENV.PY_MODULE }}
  - name: textIndexer
    uses:
      jtype: DocVectorIndexer
      with:
        index_file_name: "text.json"
      metas:
        workspace: ${{ ENV.HW_WORKDIR }}
        py_modules:
          - ${{ ENV.PY_MODULE }}
```

2. For the image input, it uses the `ImageCrafter` executor to preprocess the data, followed by the `ImageEncoder` executor to encode the image data, and finally, indexes it using `DocVectorIndexer`:

```
  - name: craftImage
    uses:
      jtype: ImageCrafter
      metas:
        workspace: ${{ ENV.HW_WORKDIR }}
        py_modules:
          - ${{ ENV.PY_MODULE }}
    needs: gateway
  - name: encodeImage
    uses:
```

```
        jtype: ImageEncoder
      metas:
        py_modules:
          - ${{ ENV.PY_MODULE }}
  - name: imageIndexer
    uses:
      jtype: DocVectorIndexer
      with:
        index_file_name: "image.json"
      metas:
        workspace: ${{ ENV.HW_WORKDIR }}
        py_modules:
          - ${{ ENV.PY_MODULE }}
```

3. It then passes the result of both `TextIndexer` and `ImageIndexer` into the `WeightedRanker` executor, which calculates the final relevance score and produces the output:

```
  - name: weightedRanker
    uses:
      jtype: WeightedRanker
      metas:
        workspace: ${{ ENV.HW_WORKDIR }}
        py_modules:
          - ${{ ENV.PY_MODULE }}
    needs: [ textIndexer, imageIndexer ]
  - name: keyvalueIndexer
    uses:
      jtype: KeyValueIndexer
      metas:
        workspace: ${{ ENV.HW_WORKDIR }}
        py_modules:
          - ${{ ENV.PY_MODULE }}
    needs: weightedRanker
```

To interact with the multimodal application in the web browser via the UI, you can use the `index.html` file provided in the `static` folder. Running the application should open the HTML file by default, but if it doesn't, then you can open the `index.html` file from the `static` folder.

Summary

In this chapter, we have covered how to put together all the components and concepts that we have learned in the previous chapters. We have walked you through the process of building basic search examples with Jina for different data types, including a text-to-text search, image-to-image search, and multimodal search, which combines both the text and the images. The things that we learned in this chapter will act as a building block for *Chapter 7, Exploring Advanced Use Cases of Jina*, where you will learn about building advanced examples using Jina.

In the next chapter, we will continue on the same journey and see how to build advanced search applications with Jina using what we have learned so far.

7

Exploring Advanced Use Cases of Jina

In this chapter, we discuss more advanced applications of the Jina neural search framework. Building on the concepts we have learned in the previous chapters, we will now look at what else we can achieve with Jina. We will examine multi-level granularity matches, querying while indexing, and a cross-modal example. These are challenging concepts in neural search and are required to achieve complex real-life applications. In particular, we will be covering these topics in this chapter:

- Introducing multi-level granularity
- Cross-modal search with images with text
- Concurrent querying and indexing data

These cover a wide variety of real-life requirements of neural search applications. Using these examples, together with the basic examples in *Chapter 6*, *Basic Practical Examples with Jina*, you can expand and improve your Jina applications to cover even more advanced usage patterns.

Technical requirements

In this chapter, we will build and execute the advanced examples provided in the GitHub repository. The code is available at `https://github.com/PacktPublishing/Neural-Search-From-Prototype-to-Production-with-Jina/tree/main/src/Chapter07`. Make sure to download this and navigate to each of the examples' respective folders when following the instructions for how to reproduce the use cases.

To run this code, you will need the following:

- macOS, Linux, or Windows with WSL2 installed. Jina does not run on native Windows.
- Python 3.7 or 3.8
- Optionally, a clean new virtual environment for each of the examples
- Docker

Introducing multi-level granularity

In this section, we will discuss how Jina can capture and leverage the hierarchical structure of real-life data. In order to follow along with the existing code, check the chapter's code for a folder named `multires-lyrics-search`. This is the example we will be referring to in this section.

This example relies on the `Document` type's capacity to hold chunks (child documents) and refer to a specific parent. Using this structure, you can compose advanced arbitrary level hierarchies of documents within documents. This mimics various real-life data-related problems. Examples could be patches of images, sentences of a paragraph, video clips of a longer movie, and so on.

See the following code for how to perform this with Jina's Document API:

```
from jina import Document
 document = Document()
chunk1 = Document(text='this is the first chunk')
chunk2 = Document(text='this is the second chunk')
document.chunks = [chunk1, chunk2]
```

This can then be chained, with multiple levels of granularity, with each chunk having its own chunks. This becomes helpful when dealing with hierarchical data structures. For more information on the `Document` data type, you can check the *Understanding Jina components* section in *Chapter 4, Learning Jina's Basics*.

In this example, the dataset will be composed of lyrics from various popular songs. In this case, the granularity is based on linguistic concepts. The top level will be the entire contents of the body of a song's lyrics. The level under that will be individual sentences extracted from the top-level body. This splitting is done using the `Sentencizer` Executor, which splits the long piece of text by looking for specific separator text tokens, such as `.` or `,`.

This application helps showcase the concept of **chunking** and its importance in search systems. This is important because, in order to get the best results in a neural search system, it is best to search with text inputs of the same length. Otherwise, the context-to-content ratio will be different between the data you are searching with and the data you have trained your model on. Once we have built the example, we can visualize how the system is matching input to output via a custom frontend.

Navigating through the code

Let's now go through the logic of the app and the functions of each component. You can follow along with the code in the repository, at `https://github.com/PacktPublishing/Neural-Search-From-Prototype-to-Production-with-Jina/tree/main/src/Chapter07/multires-lyrics-search`. I will explain the purpose and design of the main files in the folder.

app.py

This is the main entry point of the example. The user can use this script to either index (add) new data or search with their desired queries. For indexing data, this is done from the command line as follows:

```
python -t app.py index
```

Instead of providing the `index` argument, you can also provide `query` or `query_text` as arguments. `query` starts the Flow to be used by an external REST API. You can then use the custom frontend provided in the repository to connect to this. `query_text` allows the user to search directly from the command line.

When indexing, the data is sequentially read from a CSV file. We also attach relevant tag information, such as author, song name, album name, and language, for displaying metadata in the interface. Tags can also be used by the user in whatever way they need. They were discussed in the *Accessing nested attributes from tags* subsection in the *Understanding Jina components* section in *Chapter 4, Learning Jina's Basics*.

index.yml

This file defines the structure of the Flow used when indexing data (adding data). Following are the different configuration options provided in the file:

- `jtype` informs the YAML parser about the class type of this object. In this case, it's the `Flow` class. The YAML parser will then instantiate the class with the respective configuration parameters.

- `workspace` defines the default location where each Executor might want to store its data. Not all Executors require a workspace. This can be overridden by each Executor's `workspace` parameter.

- `executors` is a list that defines the processing steps in this Flow. These steps are defined by specific classes, all of which are subclasses of the `Executor` class.

The indexing Flow is represented by the following diagram:

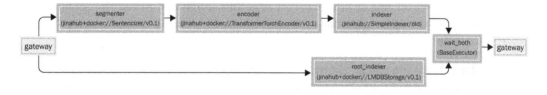

Figure 7.1 – Index Flow showing document chunking

Notice how the data Flow is split at the gateway. The original document is stored as is in `root_indexer`, for future retrieval. On the other path, the document gets processed in order to extract its chunks, encode them, and then store them in the indexer.

Following are the different Executors used in this example:

1. The first one is `segmenter`, which uses the `Sentencizer` class, from Jina Hub. We use the default configuration. This splits the body of the lyrics into sentences using a set of punctuation markers that usually delimit sentences, such as `.`, `,`, `;`, `!`. This is where the chunks are being created and assigned to their parent document, based on where these tokens are found in the text.

2. The next is `encoder`. This is the component in the Flow that transforms the sentence from text into a numeric format. The component uses the `TransformerTorchEncoder` class. It downloads the `distilbert-base-cased` model from the `Huggingface` API and uses it to encode the text itself into vectors, which can then be used for vector similarity computation. We will also define some configuration options here:

    *   `pooling_strategy: 'cls'`: This is the pooling strategy that is used by the encoder.

    *   `pretrained_model_name_or_path: distilbert-base-cased`: This is the deep learning model that is used. It is pre-trained and downloaded at the start time by the Executor.

    *   `max_length: 96`: This indicates the maximum number of characters to encode from the sentence. Sentences longer than this limit get trimmed (the extra characters are removed).

    *   `device: 'cpu'`: This configuration instructs the Executor to run on the CPU. The Executor can also be run on the GPU (with `'gpu'`).

    *   `default_traversal_paths: ['c']`: This computes the embeddings on the chunk level. This represents the hierarchy level of the sentences extracted by `segmenter`. We only encode these, as we will perform the search matching at this level only. Matching the entire body of a song's lyrics will not perform well, due to the amount of data a model needs to encode.

3. We will now deep-dive into the actual storage engine, `indexer`. For this, we use the Executor called `SimpleIndexer`, again from Jina Hub. This uses the `DocumentArrayMemmap` class from Jina, to store the data on disk, but at the same time, load it into memory for reading and writing as needed, without consuming too much memory. We define the following configuration options for it:

    *   `default_traversal_paths: ['c']`: These options configure the component to store the chunks of the documents. This has the same purpose as the previous usage of `default_traversal_paths`.

4. Next is another indexer, `root_indexer`. This is part of the specific requirements of this example. Before, at `indexer`, we stored only the chunks of the document. But, at search time, we need to also retrieve the parent document itself, in order to obtain the tags associated

with it (artist name, song name, and much more). As such, we need to store these documents somewhere. That is why we need this additional Executor. Usually, this will not be required, depending on your use case in your application. We define the following configuration options:

- `default_traversal_paths: ['r']`: We define that we will index the root level of the document (i.e., not chunk-level)

- `needs: [gateway]`: This tells the Flow to send requests in parallel, to two separate paths: one is sent to the `segmenter` and `encoder` path, and the other is sent directly to `root_indexer`, since this one does not depend on any Executor in the other path

You will have noticed an additional argument that is repeated across some of the Executors, `volumes`. This conforms to the Docker syntax for mounting a local directory in the Docker container, in order to mount the workspace in the running Docker container.

query.yml

This file defines the structure of the Flow used when querying data (searching data). This is different from the Flow configuration used at index time because the order of operations is different. Looking at the following diagram, we notice the main change is that the operations at query time are strictly sequential:

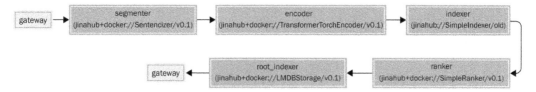

Figure 7.2 – Query Flow showing document chunking

The matches are retrieved from `indexer`, which operates at the chunk level, as we previously defined. `ranker` then creates one single match for each parent ID present in the chunks. Finally, the original metadata of this parent match document is retrieved from `root_indexer` based on its ID. This is required in order to get the full context of the chunk (the parent's full text contents and the name of the artist and song).

Just like the `index.yml` file, the `query.yml` file also defines a Flow with Executors. We will discuss their configuration choices, but we will only cover the differences from their equivalent in the `index.yml` file. If a parameter is not covered in this section, check the previous section. The following are the Executors defined in the query Flow:

- `segmenter` is the same.

- `encoder` is also the same.

- `indexer` is also the same.

5. The first new Executor is `ranker`. This performs a custom ranking and sorting of the results from the search. We use `SimpleRanker`, from Jina Hub. The only parameter here is `metric: 'cosine'`. This configures the class to use the `cosine` metric to base its ranking on. It works by aggregating the scores of a parent document's chunks (children documents) into an overall score for the parent document. This is required to ensure that the matches are sorted in a meaningful way for the client (the frontend, REST API client, or command-line interface).

6. The last hop is `root_indexer`. Here, we change `default_traversal_paths` to `['m']`. This means that we want to retrieve the metadata of the matches of the document, not of the request document itself. This takes the document's ID and performs a lookup for the metadata. As mentioned previously, `indexer` only stores the chunks of the document. We need to retrieve the full metadata of the chunks' parent Document.

Installing and running the example

I will now guide you through installing and running this example application:

1. Make sure the requirements defined at the beginning of this chapter are fulfilled.

2. Clone the Git repository from `https://github.com/PacktPublishing/Neural-Search-From-Prototype-to-Production-with-Jina` and open a terminal in the example's folder, at `src/Chapter07/multires-lyrics-search`.

3. Install the requirements:

    ```
    pip install -r requirements.txt
    ```

4. Download the full dataset. This step is optional; you can skip this step and use the sample data provided:

    I.   Begin by installing the Kaggle library if you haven't already done so. You will also need to set up your API keys as explained here: `https://github.com/Kaggle/kaggle-api#api-credentials`:

    ```
    pip install kaggle
    ```

    II. Running the following `bash` script should perform all the steps needed to download the full dataset:

    ```
    bash get_data.sh
    ```

5. The next step is to index the data. This step processes your data and stores it in the workspace of the Flow's Executors:

    ```
    python app.py -t index
    ```

6. Search your data. Here you have two options:

- `python app.py -t query_text`: This option starts a command-line application. At some point, it will ask for a phrase as input. The phrase will be processed and then used as a search query. The results will be displayed in the terminal.

- `python app.py -t query`: This starts the application in server mode. It listens for incoming requests on the REST API and responds to the client with the best matches.

In the second mode, you can use the custom frontend we have built to explore the results. You can start the frontend by running the following commands in a terminal:

```
cd static
python -m http.server --bind localhost
```

Now you can open `http://127.0.0.1:8000/` in your browser and you should see a web interface. In this interface, you can type your text in the left-side box. You will then get results on the right side. The matching chunks will be highlighted in the body of the lyrics.

Following is a screenshot of the interface:

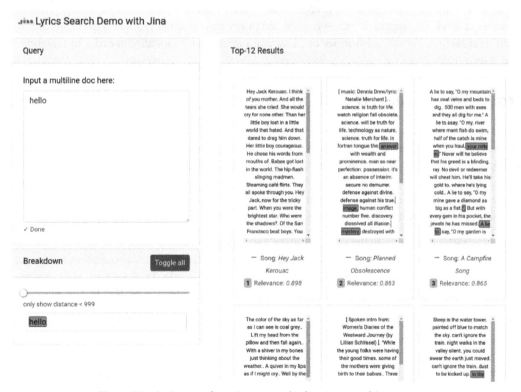

Figure 7.3 – Lyrics search engine example showing matching songs

For example, if you add the sentence `I am very happy today`, you should see a similar result. Each of these boxes you see on the right-hand side is a song in your dataset. Each highlighted sentence is a *match*. A match is a similar sentence, determined by how close two vectors are in embedding space.

Similarity can be adjusted using the breakdown slider on the left-hand side. As you move the slider to the right, you will see more matches appear. This is because we are increasing our radius in the vector space to find similar matches.

The relevance score you see at the bottom of the song box summarizes all the matches in a song. Each match has a numeric value between 0 and 1, determining how close it is to the original input in the vector space. The average of these match values is the relevance score. This means that a song with only good matches will be ranked as highly relevant.

The example also allows for more complex, multi-sentence queries. If you input two or three sentences when querying, the query Flow will break down the total input into individual "chunks." These chunks in this example are sentences, but you can determine what a chunk is for your own data when building Jina.

In this section, we have covered how you can model the hierarchical structure of real-life data in the Jina framework. We use the `Document` class and its ability to hold chunks as our representation of this data. We have then built an example application that we can use to search through song lyrics, on the sentence level. This approach can be generalized to any text (or other modality) data application. In the next section, we will see how we can leverage a document's modality in order to search for images with text.

Cross-modal search with images with text

In this section, we will cover an advanced example showcasing **cross-modal search**. Cross-modal search is a subtype of neural search, where the data we index and the data we search with belong to different modalities. This is something that is unique to neural search, as none of the traditional search technologies could easily achieve this. This is possible due to the central neural search technology: all deep learning models fundamentally transform all data types to the same shared numeric representation of a vector (the embedding extracted from a specific layer of the network).

These modalities can be represented by different data types: audio, text, video, and images. At the same time, they can also be of the same type, but of different distributions. An example of this could be searching with a paper summary and wanting to get the paper title. They are both texts, but the underlying data distribution is different. The distribution is thus a modality as well in this case.

The purpose of the example in this section is to show how the Jina framework helps us to easily perform this sort of search. We highlight how the Flow can be used to split the data processing, depending on modalities, into two pipelines of Executors. This is done with the `needs` field, which defines the previously required step of an Executor. Chaining these `needs`, we can obtain separate paths.

Let's now go through the logic of the app and what each file's purpose is. The code can be found at `https://github.com/PacktPublishing/Neural-Search-From-Prototype-to-Production-with-Jina` in the folder `src/Chapter07/cross-modal-search`.

app.py

This is the main entry point of the example. The user can call it to either **index** or **search**. It then creates the Flows and either indexes data or searches with the query from the user.

flow-index.yml

This file defines the structure of the Flow used when indexing data (adding data). I will explain the different steps.

The Flow itself has the following arguments:

- `prefetch` defines the number of documents to prefetch from the client's request.

- `workspace` defines the default location where data will be stored. This can be overridden by each Executor's `workspace` parameter.

Then, the `executors` list defines the Executors used in this Flow. Each item in this list is an Executor and its configuration.

Following is a diagram representing the indexing Flow. Notice how the path bifurcates from the gateway, depending on whether the data is image or text:

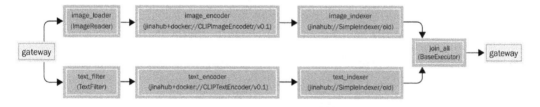

Figure 7.4 – Index Flow showing cross-modal features

We will describe the purpose of each of the Executors, grouped by paths. The first path is the path for image data:

1. The first Executor is `image_loader`. This uses the `ImageReader` class, defined locally in the `flows/executors.py` file. This will load the image files from a specific folder and pass them down further into the Flow for processing. When a document is created, we can assign it a `mime` type. This can then be used in specific Executors to perform custom logic. Here, we are using it to restrict which documents go to which Executors.

The parameters are as follows:

- `py_modules`: This tells the Python process where to find extra classes that can then be used in the `uses` parameter.

- `needs`: This creates a direct connection from the gateway (which is always the first and last hop of the Flow) to this Executor. It makes this component wait for requests from the gateway. This is required here because we want two separate paths for text and images.

2. The next one is `image_encoder`. This is where the brunt of the work is done. Encoders are the Executors that transform data into a numeric representation. It uses `CLIPImageEncoder`, version 0.1. The parameters are as follows:

- `needs`: This defines the path of the data on the image path

3. `image_indexer` is the storage for the embeddings and metadata of the documents that contain images. It uses `SimpleIndexer`. The parameters are as follows:

- `index_file_name`: This defines the folder where the data is stored

- `needs`: This makes the Executor part of the image processing path, by explicitly making it depend on `image_encoder`

4. The next elements will be part of the text path. `text_filter` is similar to `image_filter`. It reads data, but only text-based documents. The parameters used here are as follows:

- `py_modules`: This parameter again defines the files where the `TextFilterExecutor` is defined.

- `needs: gateway` defines the path of dependencies between the Executors. In this case, this Executor is at the beginning of the path and thus depends on `gateway`.

5. Next, similar to the image path, we have the encoder `text_encoder`. This processes the text and encodes it using `CLIPTextEncoder`. The parameters used here are as follows:

- `needs: text_filter`: This parameter specifies that this Executor is part of the text pat.

6. `text_indexer` stores the embeddings of the Executor.

7. Finally, we join the two paths. `join_all` joins the results from the two paths into one. The `needs` parameter here is given a list of Executor names.

You will have noticed an argument that is repeated across some of the Executors:

`volumes`: This is the Docker syntax for mounting a local directory into the Docker container.

query.yml

In this section, we will cover the query (search) Flow. This designates the process for searching the data you have indexed (stored) with the aforementioned index Flow. The configuration of the Executors is the same, at an individual level.

As can be seen from the following diagram, the Flow path is also similar. It also bifurcates at the start, depending on the data type:

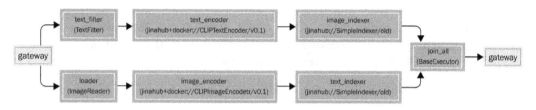

Figure 7.5 – Query Flow showing cross-modal features

The difference is that now we are searching with documents across the two modalities. Thus, `text_loader` sends the documents with text to be encoded by `text_encoder`, but the actual similarity matching is done with image documents that have been stored in `image_indexer`, from the index Flow. This is the central aspect that allows us to achieve cross-modality searching in this example.

Installing and running the example

To run the example, do the following:

1. Make sure the requirements defined at the beginning of this chapter are fulfilled.

2. Clone the code from the repository at `https://github.com/PacktPublishing/Neural-Search-From-Prototype-to-Production-with-Jina` and open a terminal in the `src/Chapter07/cross-modal-search` folder.

3. Note that this example only includes two images as a sample dataset. In order to download the entire dataset and explore the results, you will need to download it from Kaggle. You can do so by registering for a free Kaggle account. Then, set up your API token. Finally, to download the `flickr 8k` dataset, run the following command in a terminal:

    ```
    bash get_data.sh
    ```

4. To index the full dataset, run the following:

    ```
    python app.py -t index -d f8k -n 8000
    ```

5. Starting the index Flow and indexing the sample data is done from the command line, like so:

    ```
    python app.py -t index
    ```

This creates the index Flow, processes the data in the specific folder, and stores it in a local folder, `workspace`.

6.   Then, in order to start the searching Flow and allow the user to perform a search query, you can run this command:

```
python app.py -t query
```

Let's begin by running a small test query. This test query actually contains both an image and a text document. The text is the sentence `a black dog and a spotted dog are fighting`. The image is `toy-data/images/1000268201_693b08cb0e.jpg`. The system then searches with both the image and the text, in a cross-modal manner. This means the image is used to search across the text data and the text is used to search across the image data.

The text results from searching with the image will be printed in your terminal as follows:

```
text_encoder@68319[I]:   text_encoder@ 1[L]: Executor CLIPTextEncoder started
        Flow@68132[I]: 🎉 Flow is ready to use!
    🔗 Protocol:          GRPC
    🔒 Local access:      0.0.0.0:45678
    🔒 Private network:   192.168.178.144:45678
    🌐 Public address:    2001:16b8:46e2:6200:fe8:cfea:7e63:4c48:45678
Request duration: 0.20777654647827148
Searching with image toy-data/images/1000268201_693b08cb0e.jpg. Matches:
    -- text: "a child in a pink dress is climbing up a set of stairs in an entry way ." score: 0.6692,
    -- text: "a black dog and a spotted dog are fighting" score: 0.8351,
```

Figure 7.6 – Cross-modal search terminal output

The image results will be shown in a `matplotlib` figure as follows:

Best matches for 'a black dog and a spotted dog are fighting'

score=0.6858

score=0.8351

Figure 7.7 – Cross-modal search plot output

In this case, a lower score is better, as it measures the distance between the vectors.

You can pass your own image queries with the following:

```
python app.py -t query --query-image path_to_your_image
```

The `path_to_your_image` variable can be provided as either an absolute or relative path, from the terminal's current working directory path.

Or, for text, you can do it like so:

```
python app.py -t query --query-text "your text"
```

In this section, we have covered how the Jina framework allows us to easily build a cross-modal search application. This is possible due to Jina's universal and generalizable data types, mainly the document, and flexible pipeline construction process. We see that the `needs` parameter allows us to split the processing pipeline into two paths, depending on the *mime* type. In the following section, we will see how we can serve data while modifying it.

Concurrent querying and indexing data

In this section, we will present the methodology for how to continuously serve your client's requests while still being able to update, delete, or add new data to your database. This is a common requirement in the industry, but it is not trivial to achieve. The challenges here are around maintaining the vector index actualized with the most recent data, while also being able to update that data in an atomic manner, but also doing all these operations in a scalable, containerized environment. With the Jina framework, all of these challenges can be easily met and overcome.

By default, in a Jina Flow, you cannot both index data and search at the same time. This is due to the nature of the network protocol. In essence, each Executor is a single-threaded application. You can use sharding to extend the number of copies of an Executor that form an Executor group. However, this is only safe for purely parallel operations, such as encoding data. These sorts of operations do not affect the state of the Executor. On the other hand, **CRUD (Create/Read/Update/Delete)** are operations that affect the state. Generally, these are harder to parallelize in scalable systems. Thus, if you send a lot of data to index (to add) to your application, this will block all searching requests from your clients. This is, of course, highly limiting. In this solution, I will show how this can be tackled within Jina.

The key component of the solution is the **HNSWPostgresIndexer** Executor. This is an Executor for the Jina framework. It combines an in-memory HNSW vector database with a connection to a PostgreSQL database. The metadata of your documents is stored in the SQL database, while the embeddings are stored in RAM. Unlike the applications in the previous examples, it does not require two distinct Flows. All the CRUD operations are performed within one Flow life cycle. This is possible due to the Executor's capacity to synchronize the states between the SQL database and its in-memory vector database. This can be configured to be done automatically or can be triggered manually at the desired time.

Let's now delve into what each component of this example is doing. The code can be found at `https://github.com/PacktPublishing/Neural-Search-From-Prototype-to-Production-with-Jina` in the folder `/src/Chapter07/wikipedia-sentences-query-while-indexing`.

app.py

This is the main entry point of the example. The user can call it to start the index and search Flows or to search documents. In order to start the Flows, you run `app.py` as follows:

```
python app.py -t flow
```

This will initialize the Flow of the Jina application, with its Executors. It will then add new data to the **HNSWPostgreSQL** Executor, in batches of five documents at a time. This data is at first only inserted into the SQL database. This is because the SQL database is considered the primary source of data. The **HNSW** vector index will be gradually updated based on the data in the SQL database. Once there is data present, the Executor will automatically synchronize it into the HNSW vector index. This process continues until the data is fully inserted. Once one round has been completed, there will be data available for searching for the user. The user can then query the data with the following command:

```
python app.py -t client
```

Then the user will be prompted for text input for a query. This text will then be encoded and compared with the existing dataset to get the best matches. These will be printed back to the terminal.

flow.yml

This file defines the structure of the Flow used both when indexing data (adding data) and searching. I will explain the different options.

Following is the diagram of the index Flow. Notice that it is quite simple: we are just encoding and storing the encoded data. The complexity of this example application arises from the internal behavior of the **HNSWPostgreSQL** Executor.

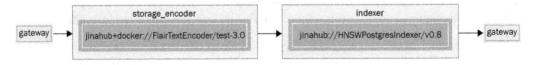

Figure 7.8 – Query Flow showing concurrency

The Flow itself has the following arguments:

- `protocol`: Defines that the Flow should open its HTTP protocol to the exterior
- `port_expose`: Defines the port for listening on

Then, the Executors define the steps in the Flow:

- The first one is `storage_encoder`. This uses `FlairTextEncoder` from Jina Hub. This encodes the text into a vector, for the linear algebra operations required in machine learning.
- The second one is `indexer`. This uses `HNSWPostgresIndexer`, also from Jina Hub. The parameters used here are the following:

  - `install_requirements`: Setting this to `True` will install the libraries required for this Executor
  - `sync_interval`: How many seconds to wait between automatically synchronizing the data from the SQL database into the vector database
  - `dim`: The dimensionality of the embeddings

You will have noticed an additional argument that is repeated across some of the Executors:

- `timeout_ready`: This defines the number of seconds to wait for an Executor to become available before it's canceled. We set it to -1 so we wait as long as it's required. Depending on your scenario, this should be adjusted. For example, if you want to safely terminate a long-running downloading request, you can set it to whatever amount of seconds you want to wait for the Executor to start.

Installing and running the example

Before running this example, make sure you understand the basic text search from the previous chapter, the chatbot example. Also, you will need to install Docker on your computer:

1. Clone the Git repository from `https://github.com/PacktPublishing/Neural-Search-From-Prototype-to-Production-with-Jina/tree/main/src/Chapter07/wikipedia-sentences-query-while-indexing` and open a terminal in the example's folder.
2. Create a new Python 3.7 environment. Although it is not required, it is strongly recommended.
3. Install the requirements:

```
pip install -r requirements.txt
```

The repository includes a small subset of the Wikipedia dataset, for quick testing. You can just use that. If you want to use the entire dataset, run `bash get_data.sh` and then modify the `DATA_FILE` constant (in `app.py`) to point to that file.

4. Then start the Flow with the following command:

```
python app.py -t flow
```

This creates the Flow and establishes the data synchronizing loop, as described in `app.py` previously.

5. In order to query the data, run the following:

```
python app.py -t client
```

You will then be prompted for some text input. Enter whatever query you wish. You will then get back the best matches for your query.

Since the Flows expose an HTTP protocol, you can query the REST API with the Jina Client, cURL, Postman, or the custom Swagger UI built within Jina. The Swagger UI can be reached at the URL informed by the Flow, in the terminal. Usually, it's `http://localhost:45678/docs`, but it depends on your configured system.

In this section, we have learned how we can use the `HNSWPostgreSQLIndexer` Executor to concurrently index and search data in our live system. In the previous examples, the Flow needed to be redefined and restarted in order to switch between the two modes. Since this Executor combines both the metadata store (via a connection to a SQL database) and the embeddings index (via an in-memory HNSW index), it is possible to perform all CRUD operations within one Flow life cycle. Using these techniques, we can have a real client-facing application that is not blocked by the need to update the underlying database in the index.

Summary

In this chapter, we have analyzed and practiced how you can use Jina's advanced features, such as chunking, modality, and the advanced `HNSWPostgreSQL` Executor, in order to tackle the most difficult goals of neural search. We implemented solutions for arbitrary hierarchical depth data representation, cross-modality searching, and non-blocking data updates. Chunking allowed us to reflect on some data's properties of having multiple levels of semantic meaning, such as sentences in a paragraph or video clips in longer films. Cross-modal searching opens up one of the main advantages of neural search – its data universality. This means that you can search with any data for any type of data, as long as you use the correct model for the data type. Finally, the `HNSWPostgreSQL` Executor allows us to build a live system where users can both search and index at the same time, with the data being kept in sync.

Index

Packt.com

Subscribe to our online digital library for full access to over 7,000 books and videos, as well as industry leading tools to help you plan your personal development and advance your career. For more information, please visit our website.

Why subscribe?

- Spend less time learning and more time coding with practical eBooks and Videos from over 4,000 industry professionals

- Improve your learning with Skill Plans built especially for you

- Get a free eBook or video every month

- Fully searchable for easy access to vital information

- Copy and paste, print, and bookmark content

Did you know that Packt offers eBook versions of every book published, with PDF and ePub files available? You can upgrade to the eBook version at packt.com and as a print book customer, you are entitled to a discount on the eBook copy. Get in touch with us at customercare@packtpub.com for more details.

At www.packt.com, you can also read a collection of free technical articles, sign up for a range of free newsletters, and receive exclusive discounts and offers on Packt books and eBooks.

Other Books You May Enjoy

If you enjoyed this book, you may be interested in these other books by Packt:

Machine Learning on Kubernetes

Faisal Masood, Ross Brigoli

ISBN: 9781803241807

- Understand the different stages of a machine learning project
- Use open source software to build a machine learning platform on Kubernetes
- Implement a complete ML project using the machine learning platform presented in this book
- Improve on your organization's collaborative journey toward machine learning
- Discover how to use the platform as a data engineer, ML engineer, or data scientist
- Find out how to apply machine learning to solve real business problems

Practical Deep Learning at Scale with MLflow

Yong Liu

ISBN: 9781803241333

- Understand MLOps and deep learning life cycle development
- Track deep learning models, code, data, parameters, and metrics
- Build, deploy, and run deep learning model pipelines anywhere
- Run hyperparameter optimization at scale to tune deep learning models
- Build production-grade multi-step deep learning inference pipelines
- Implement scalable deep learning explainability as a service
- Deploy deep learning batch and streaming inference services
- Ship practical NLP solutions from experimentation to production

Packt is searching for authors like you

If you're interested in becoming an author for Packt, please visit `authors.packtpub.com` and apply today. We have worked with thousands of developers and tech professionals, just like you, to help them share their insight with the global tech community. You can make a general application, apply for a specific hot topic that we are recruiting an author for, or submit your own idea.

Share Your Thoughts

Now you've finished *Neural Search - From Prototype to Production with Jina*, we'd love to hear your thoughts! Scan the QR code below to go straight to the Amazon review page for this book and share your feedback or leave a review on the site that you purchased it from.

`https://packt.link/r/1801816824`

Your review is important to us and the tech community and will help us make sure we're delivering excellent quality content.

www.ingramcontent.com/pod-product-compliance
Lightning Source LLC
Chambersburg PA
CBHW060132060326
40690CB00018B/3846